D0794184

This text is printed on acid-free paper.

Published by John Wiley & Sons, Inc.

Library of Congress Cataloging-in-Publication Data:

Sagdeev, R. Z.
The making of a Soviet scientist : my adventures in nuclear fusion and space from Stalin to Star Wars / Roald Z. Sagdeev : edited by Susan Eisenhower.
p. cm.
Includes index.
ISBN 0-471-02031-1
1. Sagdeev, R. Z. 2. Physicists—Russia (Federation)—Biography. I. Eisenhower, Susan. II. Title.
QC16.S24A3 1994
530'.92—dc20
[B] 93-40709

Printed in the United States of America

10 9 8 7 6 5 4 3 2 1

To my wife, Susan Eisenhower,
who was brave enough to marry me
when I was still a Soviet scientist

Acknowledgments

The very idea to write this book on a now "extinct species" was suggested by my wife, Susan Eisenhower. Her constant practical help in brainstorming, writing, and editing finally made the project real. The early work on a first draft started in the end of 1988, while I was still living the life of a Soviet scientist in Moscow and could benefit from innumerable conversations with my friends and colleagues there. There is no need to mention their names here: they are the heroes of my book and appear in its pages.

I feel very indebted to many of my comrades-in-arms in the international space science community: Drs. Thomas Donahue, Roger Bonnet, Reimar Lüst, Gerhard Haerendel, Ian Axford, Renee Pellat, Louis Friedman, and many more of those who remained friends and stayed in touch even in the dark years of the Cold War.

From the beginning of 1990 work on the manuscript was continued in the United States because I had moved to Maryland (it was not the result of brain drain, but rather of a "heart drain"). A distant perspective and the interest and support of my American friends (among them I would like to mention Dr. S. C. Liu and Dr. N. Papadopoulos) were of great help. Several people assisted in the technical preparation of different parts of the draft text: Mary Ridgell and Heidi Araya.

The critical countdown for the book started when John Wiley & Sons, Inc. decided to move to the launch phase under the disciplining and friendly leadership of my editor, Hana Lane. The members of her team, Neil McAleer, Scott Renschler, and copy editor Jude Patterson, finally made it ready for printing.

Roald Sagdeev
November 1993

Contents

➢

Foreword

➢

Carl Sagan

Roald Sagdeev is an expert in plasma physics, the behavior of electrically charged gases. At first sight this science seems abstract, obscure, impractical. But things are connected. Plasma physics is related to nuclear weapons and nuclear fusion (a possible source of unlimited energy). Governments need wise counsel on these matters. Plasmas also permeate space and surround planets, which led Sagdeev to the directorship of the Soviet institute in charge of robotic exploration of the planets. Since space exploration is also a tool of national policy, Sagdeev was, by this route, led to a firsthand encounter with politics at the highest level in the old Soviet Union. And from there it was only a step to his most recent incarnation as professor of physics at the University of Maryland, becoming an effective advocate of closer cooperation between the United States and Russia, and, with fitting symbolism of the unexpectedness of recent world events, the husband of Susan Eisenhower, the granddaughter of President Dwight D. Eisenhower.

It has been my pleasure to have known Roald Sagdeev in each of his incarnations. I've witnessed him in action in the U.S. Capitol and in the Kremlin, on matters of science and strategic policy, in the presence of the powerful and the powerless. He is a remarkable person. His management of the Soviet Academy of Sciences' Space Research Institute (IKI) in Moscow was without precedent. He pioneered the involvement of scientists of fifteen nations, from East and West together, working openly and efficiently on ambitious spacecraft missions to Venus, to Halley's comet, to Mars and its moons.

He instituted glasnost before Gorbachev. In earlier times, foreign visitors entering IKI would be subject to close and unfriendly scrutiny by leather-coated KGB operatives; we were required to be accompanied everywhere, even to the bathroom. After Sagdeev, foreign visitors were recognized individually at a special entrance, warmly ushered in, and were free to go everywhere. Lost in the recesses of IKI, I once blundered into the mainframe computer area, at a moment when IKI's computer was arguably the most advanced in the Soviet Union. No one stopped me. You could even sometimes hear rock-and-roll music in the halls of IKI during Sagdeev's tenure as director. His management style was made clearest to me by the affectionate regard borne to him by the waitresses in IKI's dining room.

In a society where institute directors tended to rule for life, Sagdeev publicly proposed term limits and soon quit his prestigious position to show it could be done. He refused election to the presidium of the Soviet Academy of Sciences in favor of the celebrated dissident physicist Andrei Sakharov. When official Soviet propaganda passed on a claim that the AIDS pandemic was produced by American germ warfare laboratories, Sagdeev publicly debunked the idea. It is amazing to me that he was able to accomplish so much and advocate positions so contrary to the official Soviet line without serious personal penalty. His brilliance, organizational skills, extraordinary sense of humor, and great personal charm may be part of the answer. I think his minority status as a Tatar helped him to be critical of the system under which he was raised. Unlike so many of us, Sagdeev has not fallen into the trap of thinking that one or another of the great contending world political systems was exclusively right, much less perfect.

I remember him at a meeting on strategic weapons policy involving scientists and religious leaders in Bellagio, Italy, on the shores of Lake Como. There he renewed his acquaintance with the former president of Notre Dame University, Father Theodore Hesburgh, and like so many of us was taken by Hesburgh's charisma. At dinner one night Sagdeev rose, with his arm around Hesburgh's shoulders, to make a toast. It went something like this: "Being raised in a Communist society, I was taught that religion is the opiate of the masses. Having spent some time with Father Hesburgh, I think, 'I'd like a little more of that opium!'" This position would be easier to sustain were most religious leaders like Hesburgh. But the incident struck me especially because Sagdeev's remarks were so at variance with official Soviet ideology.

With a few other highly placed Soviet scientists who were also granted routine access to the highest levels of power, Sagdeev played a vital role in passing on facts, policies, and potential strategies that would otherwise have been filtered out. He played a heroic role in restraining Soviet counters to the American Strategic Defense Initiative ("Star Wars"), helping to forestall an acceleration of the nuclear arms race both in space and on Earth. Events have borne out his judgment.

This autobiography casts light on little-known and very important corners of the former Soviet Union. It reflects the author's lucidity, humor, and sober judgment. It is a fair reflection of an extraordinary man and an extraordinary career. I wish there were more people like him at the right hand of every national leader.

1 / Tickets to Mars

➢

I was born in the dark days of winter 1932. It was a time that Joseph Stalin would call the "great turning point"—his accession to power. To reach his goal of unlimited control over the nation required the elimination of the kulaks, the only remaining class that opposed the new communist system. They were wealthy farmers who produced the bulk of the nation's foodstuffs. Despite their contribution, Stalin regarded their deportation and persecution as a revolutionary necessity.

As ironic as it might sound, now that the world knows Stalin's true legacy, the early thirties was the era of grand vision for most of the Soviet people. At the time of my birth, fifteen years after the revolution, the country had embarked on a mission of historic proportions. Those in power aimed at nothing less than a new paradigm based on principles that would radically change the very nature of society itself, the principles of "scientific communism."

My parents were part of the new Soviet breed: educated academics sprung from the cradle of the peasantry. Like many other Soviet couples of their generation, they were consumed by the fervor of a time that promised limitless horizons.

Their aspirations found expression in me, and later my brothers. I was the first of their sons, named after Roald Amundsen, the famous Norwegian polar explorer who was the first man to reach the South Pole, in 1911.

My brother, who was born four years after me, was named after Robert Peary, the explorer believed to be first to reach the North Pole. Later, when two more brothers were born, my parents simply gave them Tatar names,[1] in keeping with the tradition of the village Tatars from the Volga River region, from whom we were descended.

My first memories are of Moscow, the city in which I was born. My father was studying at Moscow State University at the time. As a result

1. My parents explained that they had simply run out of poles.

he was entitled to housing in one of the university dormitories. We lived in an old building, a prerevolutionary house on the Old Arbat. Because it was the married students' dormitory, we were given a small single room for the three of us. The dormitory was configured much like a hotel. Each room's door opened out onto a long, almost endless corridor that led to the so-called conveniences, the communal bathroom and the kitchen.

When running to the kitchen to ask my mother something, I would immediately encounter the unbelievable mixture of the many smells coming from several pots simultaneously boiling on the *kerosinkas*. In one corner of the steamy room was a primitive radio from which the black paper–covered loudspeakers blared patriotic marches and other stout-hearted music. Occasionally the music would be interrupted by exciting stories and updates not only about polar explorers, but also about the great heroes of aviation. This radio was my first herald of the life surrounding us in the outside world.

The state clearly encouraged analogies linking the natural Russian impulse for exploration and adventure to ideological imperatives. It was only later that I was able to understand these analogies in terms of Stalin's real motives for promoting such romanticism and high spirits. He did so under the slogan of "the conquest of nature." These tales diverted our attention from the painful reality of our miserable national life. The country was still unable to recover from the long, first world war and the bloody internal civil war. And there were also acute conflicts that caused quickly escalating internal oppression.

In addition, Stalin used this propaganda to set all of his plans in motion on a grandiose scale. This is why the brave airplane pilots, polar explorers, and engineers inventing their miraculous machines were promoted as heroes of the epoch.

Although I was a small boy, I was fascinated by what was going on around me. The Russian ship *Cheluskin* set out on an expedition with more than a hundred people to prove that the Northern Sea (the Arctic Ocean path) could be used for navigation. However, in an attempt to navigate through the Bering Strait to the Sea of Okhotsk, their ship was imprisoned by the ice. Apparently, in preparing for such an expedition, they had not developed sufficient technological advancements. Instead of using an ice breaker, they had used a conventional cruise ship, which in no time had found itself impaled by ice. The ship sank, leaving its crew and expedition in dangerous Arctic ice fields. Despite the catastrophe, our propaganda used even this event to paint an heroic story of the rescue of the *Cheluskin* expedition. The *Cheluskin* expedition was finally saved from the ice by airplanes that landed on a tiny spot on the iceberg.

I remember vividly the heroes of this operation. The head of the expedition, Otto Schmidt, rapidly became a famous Soviet scientist of the time. Later Schmidt was more and more involved in other prestigious adventures, until he eventually became one of the model Soviet scientists promoted by Stalin as an example of the new breed of Soviet intelligentsia.

The whole country was challenged to establish records in everything. A kind of record mania was rampant: records to fly the longest distance in a plane; records to reach the highest altitudes in a plane or stratospheric balloon; and records to excel in predominantly physical jobs such as coal mining, despite sacrifices and failures. For this last-named type of record, those given the title Hero of Socialist Labor were decorated and promoted as exemplary workers. They were called Stakhanovites, named for coal miner Alexei Stakhanov who, during one working day, produced enough coal to fulfill almost twenty quotas.

To inspire people to such undertakings, the state promoted an almost mystical approach to new technology and new machines. Airplanes were never called simply airplanes but rather steel birds. Tractors were called steel horses, and the whole country was compared to a giant machine with every man playing the role of the nuts and bolts. Society was a fast-moving train speeding toward communism.

As kids we enthusiastically sang the songs of the times, composed to produce an aura of advancement and to promote an obsession with machines, as well as exploration and adventure. We had songs such as: "The mind gave us steel arms-wings, and instead of a heart, a fervent motor."

The airplane pilots of that period were probably the most prestigious, most beloved breed in the whole nation. In our press, they were called Stalin's Falcons. It is not surprising that many first-generation Soviet apparatchiks wanted their sons to become airmen. Stalin's son Vasily was a typical example, becoming an airman and, eventually, one of the air force's chief commanders.

In many ways, we children of the new Soviet breed of intelligentsia were part of the grandiose special experiment sponsored by Stalin and the

Communist party, many of whose ideas went back to Vladimir Lenin. The experiment called for the gradual replacement of the old intellectuals, who were considered spoiled by the bourgeois mentality. They were to be replaced with a new breed of intellectuals who would be elevated from the bottom of the social strata, the working class. From the party's standpoint, they would be completely controllable.

My father was promoted with this political imperative in mind. He came from a remote rural area in what is now known as Tatarstan. An ethnic Tatar, he worked hard and was academically inclined. At the age of fifteen, he was sent to a high school outside his village where dozens of young boys and girls, chosen for their promising capabilities, were brought for intensive studies. After completing his studies at this special school, he was sent to Moscow to study mathematics at the university.

I heard from my father many stories about these people who were professors and teachers in his classes at the university. From my earliest memories, the name Igor Tamm, one of the physicists who lectured for my father's class, was a prominent figure in his stories.

Even though my father and his classmates were unsophisticated and inexperienced, these young boys—the offspring of workers and peasants—gradually could see the huge difference between the professors capable of exciting them with real knowledge and the so-called Red professors, capable only of preaching the quasi-philosophical quotations from Karl Marx, Lenin, and Stalin.

At the time, I had no idea that one of my great heroes, even idols, Konstantin Tsiolkovsky, who is known as the father of Soviet rocketry, was also "co-opted" by the regime for their own purposes. Tsiolkovsky's life was an inspiration for many young kids of my generation. In his early years—during the late nineteenth century—he worked as a schoolteacher, deep in a rural area. He was a self-made man; he even received his educational diploma in an unconventional way, passing all the examinations without ever attending college. He had difficulty communicating with the external world; scarlet fever during his early childhood had left him almost completely deaf. Perhaps that is the reason why he constructed his own fantasy world, in which he created models of future flying machines and formulated the principles of rocketry.

Almost unknown in prerevolutionary Russia—though supported and recognized by geniuses of Russian science like Dmitri Mendeleev, author of the periodic table of the elements, and Nikolai Zhukovsky, founder of Russian aerodynamics—Tsiolkovsky laid the foundation for the future success of Soviet aviation.

Tsiolkovsky's earliest fascination was with the flying machines and stratospheric balloons of the late nineteenth century and the prototypes of twentieth-century dirigibles. He wrote several articles and sent them to the Academy of Sciences. At that time, he understood the limits of flying based on the interaction between the machine and air.

Tsiolkovsky, a self-taught scientist, forced himself through the levels of mathematics and mechanics needed to write down and to derive the basic principles of rocketry. This body of work is still used in textbooks and is sometimes called the Tsiolkovsky equation. In scientific terms, this equation was his greatest achievement. Since he was and remained, until his death, an amateur and a dreamer, it is remarkable to imagine him forced to live a routine life as a poor, struggling provincial schoolteacher, constantly immersed in an imaginary world of technical inventions and dreams. He invented the principle of multistage rockets, and he considered constructions of orbital spacecraft—the prototypes of future space stations. He discussed different scenarios for docking in space and landing on earth. One can even find in his writings a description of something that can be classified as a shuttle.[2]

Stalin used Tsiolkovsky to further the idea of the superiority of the socialist system over the capitalist one. In 1935 Tsiolkovsky was invited to perform a very special task: to preach from the most prestigious podium in Red Square during the May Day demonstration. There he delivered a message about the future space adventures of the Soviet people. He talked about the impressive air parade that was taking place above Red Square, calling the airplanes steel dragonflies. The dirigibles flying low over Red Square were the materialized dream of his youth. For these achievements, he praised the Communist party.

2. While professional scientists were sympathetic to Tsiolkovsky, they never treated him as their equal. It is remarkable how many of his dreams and predictions were finally realized. In his last resting place, the city of Kaluga, where he spent many years of his life teaching in the local school, there is a museum named after him. Among the various exhibits, visitors can find the German edition of Tsiolkovsky's book on space flight and rocketry, which was taken as a trophy by the Soviet army after capturing Penemunde, the secret place where the Nazis were developing the V-2 rocket under the leadership of Wernher von Braun. This particular copy of the Tsiolkovsky book is especially meaningful because on almost every page one can see the pencil marks made by von Braun. It proves that practical scientists and engineers found a lot of interesting ideas, hints, and inspiration in the early works of Tsiolkovsky.

Enthusiastic young people and elevated peasants and workers walked through Red Square, my parents among them. The voice of Tsiolkovsky told them about future interplanetary flights. He said that he had spent forty years of his life developing the ideas of rocket engines and had always thought that flights to Mars would start only after many centuries. But now he reconsidered his forecast. Addressing the crowds, he said that many of them would have the chance to witness the first extra-atmospheric voyages.[3]

A few months after his Red Square speech, Tsiolkovsky died. On a bleak autumn day, the funeral cortege was carried through the streets of his native Kaluga. Tens of thousands of people jammed the way, straining to pay homage to this simple, reclusive schoolteacher from deep within the Russian countryside. The funeral procession, one of the largest in our recent history, represented just the kind of worship required to promote the country's theocratic state.

Tsiolkovsky, believed by some to be a prophet and a saint, seemed to link the peasants and the countryside—the primitive aspect of traditional Russian life—with an image of the shining, sophisticated future. Somehow both worlds existed compatibly in Russia.

This reminds me of a scene from an early Soviet science fiction satire popular in the twenties. In the midst of the poor, semirural landscape of Russia, only a few years after the civil war, two enterprises are competing against one another from wooden booths. One of them sells sausage. The other sells tickets to Mars.

3. Tsiolkovsky prophesied accurately. His predictions came true only twenty-two years later, in 1957, when *Sputnik I* was launched by a Soviet rocket.

2 / Going to Kazan

➢

The new Soviet state touched the life of every individual in a most profound way; it also dramatically shaped the histories of countless families. Mine was no exception. My father, catapulted from his Tatar village to life in the capital of the Soviet Union, completed his undergraduate studies at the world-famous Moscow State University. On one of his summer vacation trips back to his native village, he married a young village girl, Fakhria Idrisova, and brought her back to Moscow. A little more than a year later I was born.

After graduation my father was given a fellowship for postgraduate work on his doctorate in mathematics. Like many of his contemporaries, he fell victim to the general obsession with political activism. My father's political duty was to promote trade unions among the scientific workers and academics. He became one of the leaders of this effort.

When I was about four years old, my parents unexpectedly decided to leave Moscow to live permanently in Kazan, the capital of the Tatar Autonomous Soviet Socialist Republic on the Volga River. During our last year in Moscow, my father's health had been steadily deteriorating. During a routine checkup, an incompetent doctor told him that something was wrong with his heart. The news dealt a very strong blow to Papa, who had distinguished himself as an accomplished native Tatar wrestler. Because of it, I think, he developed a psychological phobia that at any second his heart might stop beating. Such a prospect led him to severe bouts of depression.

If it is true that an awakening of consciousness is always associated with great emotional shock, then, in my case, the first signal of my self-consciousness came with a mortality shock. It probably came in conjunction with my father's syndrome. Somehow it entered my mind that life has a finite span. I was terrified, really shaken, by this revelation. I could not have found a better person with whom to share my terrifying discovery than my father, who was himself a patient of mortality melancholia.

So I asked questions: "Father, is it true? Is it true that we are going to die eventually?" Later on I understood that the mortality shock I experienced—though universal—is probably expressed very uniquely in an atheistic society. Ours was a society that, on the surface at least, denied the existence of a divine presence or a soul. My father tried to calm me with stories. First, he tried to divert my attention from these deep, eternal, philosophical questions. But I was persistent: I remember that these thoughts occupied my mind for quite a long period of time. Father would try to soothe me by saying: "Well, you know, by the time you're grown up, science will have found a way to make human beings immortal. Look what is happening now. Look at how actively scientists are engaged in solving such problems."

He referred to our recent visit to the mausoleum in Red Square where, for the first time, I saw the shell of Grandfather Lenin. My father said, "This is the miracle that was provided by science." I was, of course, capable of understanding that it was only a half-miracle because Lenin was unable to talk and move his arms and legs, and he was lying speechless. But at least it was material proof of the very first successes of science.[1]

I did not really trust the idea that science would be quick and efficient enough to promote immortality during my lifetime. However, some by-product of these philosophical discussions with my father may have given me the first impulse to become a scientist.

In the summer of 1937, my parents took me with them by train to our new home, the city of Kazan, capital of the Tatar Autonomous Republic. For my parents, particularly my father, it could not have been easy to

1. In retrospect, Lenin's body was less a triumph of "science" than of politics. Later I understood why such achievements were important for Stalin. Stalin tried to keep Lenin's body hostage to promote his own policies in the country, to keep the population engaged in the sacrosanct essence of communist ideas. As a matter of fact, to preserve the body of Granddaddy Lenin, they had to establish an entire specialized institution. The leading scientific director of this institute was eventually promoted to the Soviet Medical Academy. The second windfall of his career was to perform the same type of sacrosanct procedure on Stalin's body. However, it most likely gave him only very brief satisfaction because, as a result of Khrushchev's speech at the Twentieth Party Congress only a few years after Stalin's death, Stalin's mummy was taken out of the mausoleum. Lonely Lenin was all that was left. All alone . . . but now nobody knows for how long.

leave Moscow and return to the Tatar Republic, to "benefit" from the quieter life. My father accepted a position as a lecturer of mathematics at the local pedagogical institute in Kazan. My mother went back to teaching as well. Again we made our home in a dormitory for both students and teachers.

We returned to the country where my Tatar ancestors had lived for many centuries. The picturesque city of Kazan—a mixture of European and Oriental styles—lies on the eastern side of the Volga River, 500 miles from Moscow.

My father's family still lived in a small Tatar village of wooden huts. My grandfather was a strong man. I remember his loud, booming voice, which sometimes forced me to escape to the cozy corner of the hayloft in the backyard of his modest home. Despite his strength, however, he lived with a childhood affliction that had damaged his feet. Walking was very difficult for him. He served as the village postmaster from his home in a central spot in the village, sorting the mail and talking to the steady stream of villagers who would come to collect their mail and newspapers. Often he would get his children or grandchildren to carry the mail that required delivery.

My grandmother, too, was quite well known in their village. She was said to have had some special skills in health care and medical matters. The village had no doctor or any kind of medical personnel, so my grandmother was called upon to assist with the occasional birth or, as my parents would tell me proudly, to serve as the village dentist. Apparently she was particularly good at pulling teeth. I don't know whether or not she had any special medical abilities, but it is certain that she did not have even the slightest formal education.

Though the village was situated only five miles from the Volga, the main transportation artery for the country, electricity had not yet arrived, although it was expected to do so, according to the grandiose plan of electrification known as Lenin's Plan of the early twenties. According to Lenin's vision, the combination of electrification and Soviet power would immediately mean the coming of the communist era. It was easy to understand why communism had not yet reached us. But everyone at a distance from Moscow was excited that very soon villagers would be getting electric light bulbs in their homes. For some reason, these bulbs were also called "bulbs of Ilyich."[2] I had not yet been "officially" educated as a scientist, and so I matter-of-factly explained to my grand-

2. After Lenin's patronymic.

parents that such "lamps" were invented by Thomas Edison, not by Lenin. Nevertheless, the expected arrival of the lamps in every household was regarded in an almost mystical way: as the icons of a new way of life.

My grandparents came from a traditional Muslim community, but if they were very religious, it had no impact on me. There was very little material evidence of their adherence to Islam. There was no mosque in the village and, in general, it was difficult to find any mosques in the whole Tatar Republic. Most of them had been swept away after the revolution.

As for my parents, I did not witness any interaction between them and their religion. Most likely urbanization had broken any ties they had with their early religious upbringing. The only story I can recall took place somewhat later, in the late forties after the war. By that time my father held the position of deputy to the prime minister for cultural affairs. In that capacity he was severely criticized by authorities for granting permission to his mother to invite a Muslim holy man, from a distant part of the Tatar Republic, to come and read the Koran at my grandfather's funeral.

My grandparents, both of whom spoke no Russian, were surprised and, I think, shocked that I, their eldest grandchild, was unable to talk to them in Tatar. I was immediately branded "the little Russian," which was insult enough to serve as the stimulus I needed to learn Tatar as fast as I could.

I understood from a very early age that all of humankind was divided into two large groups: ethnic Russians and ethnic Tatars. In cases of conflict, each group would use the occasion to remind the other about the difference in ethnic origin. It was a way of life.

By today's standards, it seems strange that I was not raised from the beginning to be bilingual, but it was quite common in the twenties and thirties for parents of Tatar origin to teach their children only Russian. Among the generation having children at that time, there was a strong urge to be integrated into Russian culture.

Frankly speaking, I had no feeling of protest against Russification at that time, although I am now fully conscious of what it has meant for my native land. At that time, I thought nothing of it. I was absolutely charmed with the richness of the Russian language, and every new book would bring me further proof of the beauty of Russian literature. In a Russian-speaking environment, I would feel ashamed to respond to a question addressed to me in Tatar. I took my Tatar language, which was far from perfect, as inferior—its only purpose being to communicate with my relatives who didn't speak Russian at all.

I had the privilege of being part of two different circles among children. In summertime while in the village, I was among my little Tatar compatriots, with whom I was quickly able to communicate. We were completely integrated into the world of nature, full of funny animals and herbs. Going back to the city, I was immediately immersed in the world of town boys, playing very different games—games that were, in many ways, our interpretations of the world of adults. We played heroes of the civil war, fighting Reds against Whites. Nobody wanted to play Whites. It was a form of "big punishment" if someone was eventually delegated to represent our class enemies.

Even as small children we were aware that something was going on in our society. I remember that not far from our house there was a cottage guarded by the militia. We knew that it was the house of one of the biggest bosses in the whole area, in fact the home of the first secretary of the regional party in the Tatar Republic. It was a noticeable event when we were told one day that, unfortunately, this particular first secretary had turned out to be an "enemy of the people." His house was abandoned for quite a while before someone else was named first secretary and moved in.

I remember, before the war, my parents would say that a certain acquaintance had been arrested. They would try to speak in very low voices to avoid getting me excited about the subject. But I had the definite impression that they were depressed when speaking about such things.

I asked more serious questions a few years later when I went to school. Because of the war, there were terrible difficulties finding the textbooks for every discipline we studied. For that reason, we would be issued used textbooks that had been printed a few years before. I can still vividly remember the striking sensation of reading one chapter after another until coming across the portraits of leaders of our country, all the way up to the Politburo. We would be instructed by our teacher to cross out certain names and blacken their pictures. Later, "enemy of the people" would be written on the page. It was rather difficult to get rid of the feeling that there were too many enemies who were originally given a place of prominence in our history and in our textbooks.

I think my father, a party member, was perhaps spared some dreadful fate because of his illness. He was not particularly active in party matters and seemed to bounce from one extreme to another—from hospital to home to hospital again.

When he was home, he was completely occupied by his mathematical teachings. Occasionally I would accompany him to his lectures at his pedagogical institute. I felt that if I accompanied him to his lectures, I would give him some additional moral strength. And it made me very proud to be able to help him fight his illness in this way. These outings were also some of the most exciting moments of my childhood. It was during such moments that what seemed to be hidden areas—the special studios, the cabinets, the mysterious collections of scientific instruments that were used to teach students physics and mathematics—were brought to life for me. Though the mysteries of science were as yet unexplainable to me, this seemed to be a world of harmony and fascination. What a joy it was to wear bicolored glasses with red and green filters and watch two-plane drawings suddenly become stereoscopic reincarnations of three-dimensional geometry.

How sad it was to return to earth, where everyone was called upon to be ever vigilant in facing the internal enemies. We kids, too, tried to be on the alert, reflecting the mood of the adults. But we couldn't comprehend who these enemies were who opposed the idea of socialism. Where were these remnants of capitalism?

We knew that even if our country were alone, an isolated island of socialism in the dangerous sea of capitalism, Comrade Stalin had proved that it could survive and grow. We knew that at any moment the enemy could be kicked out by the powerful, ever-ready Red Army. In that, we were completely reassured by victorious songs like: "With roaring fire and flashing flight of steel, tank men will carry out violent attacks, when orders come to go from Comrade Stalin." The songs could be heard coming from everywhere, from loudspeakers and from crowds demonstrating on important national festivals.

No one could doubt that it must be true since, in Russian, *steel* and *Stalin* rhymed in these songs.

3 / War

➢

Early one morning just after my first school year was over, the radio announced terrifying news about the treacherous surprise attack by the Nazis along the entire Soviet-German border. June 22, 1941—I will never forget that somber day.

The radio broadcast a speech by Vyacheslav Molotov, who was prime minister of the country at that time. There was no way to think that it was going to be an easy war. It was very clear that the country would have to go through a period of severe trials.

The communiqué from the front lines during the next few days created a real shock for me and every child of my generation. Instead of inflicting victorious punishment upon the aggressor, the Red Army fled from the Germans in a state of complete panic. The radio blared: "Retreating under the pressure of the greater, superior forces of the Fascist army, heroic Soviet soldiers were abandoning cities. . . ."[1]

We couldn't understand what was happening. Where were the invincible forces of the Soviet tank men? My first naive thought was that maybe, for some very secret and superior consideration, Comrade Stalin had decided not to give the order immediately—the order to crush enemy forces. Indeed, we were worried about what had happened. Where was his familiar voice on the radio, with its pronounced Georgian accent? Instead we heard only an appeal from Molotov; Stalin's silence continued for almost two weeks.

Though no one tried to be outspoken, I remember that the adults conveyed their concern. Some of them even expressed disapproval of the long silence that came from the great Father of the People. Many years later, the historians of World War II would tell us that Stalin was in a state of complete panic, unable to come to his wits during the first days of the war. One can only guess at the principal explanation: mortal fear of potential

1. As if commenting on such official releases, famous Soviet poet Aleksandr Tvardovsky later said, "Cities are abandoned by soldiers, but always taken by generals."

defeat, or extreme humiliation by the treacherous Hitler, who had broken the nonaggression treaty, the cornerstone of Stalin's prewar policy.

Whatever the explanation, Stalin's lack of action cost the nation dearly. The flower of the army, the bright officers and commanders, had been a special target of Stalin's purges a few years prior to the war. Stalin's first marshals, those spared to lead the tank armies, were in fact older men who had been prominent during the cavalry battles of the civil war, 1918 to 1920. They were still great, riding on horses during Red Square military parades. However, the front lines rolled menacingly fast toward Moscow, Leningrad, and other key cities of the nation. Mass mobilization of men into the army left almost every family without fathers, uncles, and even grandfathers.

My father was left untouched because of his medical problems. It was the source of a deep inferiority or guilt complex that I felt during the first years of the war. While the fathers of my contemporaries went to defend us against the Fascists, I was unable to reconcile their absence with the presence of my father at home. For consolation I tried to count how many of my uncles, my closest relatives, were taken into the army. When I counted beyond a dozen I stopped this exercise. Poor kid. I did not know then that the countdown of men returning from the war would be much more painful.

There was no consolation during brief radio communiqués from the front. The part of the country falling under Nazi occupation was growing very fast, while the influx of war refugees reaching remote places like Kazan was also growing. Almost every family had to provide accommodation and hospitality for such people, ours not excluded. I remember very well that for two years one of the rooms in our house was given to a couple fleeing from Dnepropetrovsk in the Ukraine. It was probably the first time in my life that I met real dissidents. They were so shocked by what they had seen—the bombing, the seemingly invincible offense by the German troops—that they had lost not only respect for the Red Army, but respect for Soviet power in general.

Kazan was still rather far from the line. Even at the peak of the German invasion, during the late autumn of 1941 when the armies were very close to the outskirts of Moscow, they were still 450 miles from Kazan. Despite these distances, however, people were given instructions to mask their windows at nighttime so that no light could be seen. We were also told to reinforce the windows with paper plaster. But the Germans never sent bombing raids over Kazan. A couple of times we heard air defense guns firing, but later we were told that they were firing at German air reconnaissance planes.

My closest association with the potential defense of Kazan and the surrounding region was when my father took me for a two-week trip with his colleagues from the pedagogical institute. Those who were not taken into the army were sent to dig a defense trench around the city of Kazan. It was our Kazan Maginot Line. Fortunately, it was never called into service, but it did say something about the state of our preparedness. We expected that the Germans would soon reach such deep areas inside our country.

For my father, too, the war had quite a psychological impact—particularly because he was never asked to enlist. This, I believe, was responsible for curing him of his medical problems. Since very few men were left in such deep areas of the country, he was called upon to take on more and more responsibility. He was in charge of a number of campaigns, such as digging these defense lines. On another memorable occasion, I remember very well that he was sent to cut trees in the forest some sixty miles from Kazan to create a fuel stock for the city during the coming winter. I was invited to accompany him on this expedition. Our team consisted of my father, myself, and quite a number of elderly women.

In addition to the help I offered my father and his team, every child of my generation tried to contribute to the fight against our enemies, the Fascists. In the very first days of the war, I remember the country was overwhelmed with slogans warning us about German spies who were supposedly everywhere. A group of my friends and I—young boys about the age of eight or nine—formed a special voluntary brigade and tried to follow every strange person we saw in our city. We thought we were fulfilling a very important duty, but nothing came out of our spy hunt. Before long we lost interest in our fading pursuit.

The impact of war on us was strong, although indirect. We saw a thousand wounded soldiers brought to Kazan for medical treatment. Improvised hospitals were opened almost everywhere. They were given the school buildings and institutes, including my father's pedagogical institute. In the biggest auditorium, where my father used to give his lectures, there were beds everywhere for hundreds of wounded men. The sacred, dusty smell of the science temple was replaced by the smell of medicine, perspiration, and death.

We were perpetually fascinated by the big airfield on the outskirts of the city, where they brought the wreckage of warplanes. It was like a giant warplane cemetery. If you were clever enough, you could outwit the guards and enter. It was a real paradise for youngsters. Like archaeologists, we could dig and discover all sorts of disintegrated hardware. From time to time, we were delighted to find such interesting things as a radio or a compass or a watch. Occasionally we would see the wreckage not only of German planes, but also of Soviet and American ones—planes that had been brought to the Soviet Union as military aid from the United States.

Perhaps among the most important developments of the war, within the realm of my own life, was the evacuation of the Academy of Sciences from Moscow to Kazan. Evacuated enterprises, as well as scientists and their families, were brought on special trains. As there were no apartments, and virtually no accommodations that could be found immediately, the convention hall at Kazan University was converted into a dormitory for these famous, world-class scientists. To do this, they put ropes in two directions, forming a net from which blanket partitions were hung. The result was a huge number of blanket-divided cells in which whole families lived.

For weeks the best scientific brains in the country—those who were not already relegated to prisons or gulags—were living right there in the university's convention hall. For me, this was an enormously exciting development. All of us young boys were obsessed with the presence of these academicians in our city, particularly Schmidt. Schmidt had been sent to Kazan to lead the team of Soviet scientists during this evacuation period. As one of the Soviet Union's leading polar explorers and scientists, Schmidt had been one of my earliest childhood heroes.

Not long after the academy set up its headquarters in Kazan, my father and I were walking down one of the streets when we saw Schmidt and stopped him. I was introduced to this heroic figure. Standing next to me, he was curious-looking, with his long, black, curly beard. I remember my father saying that this outstanding man was destined to become the president of the academy, but Father was wrong. As it turned out, it would have been politically unacceptable to have a person of ethnic German origin as president of the Soviet Academy of Sciences, especially since we were at war against Germany. As it probably could have been predicted, he was "released from his duties" in 1942.

Schmidt was lucky; he was allowed to remain a free man. But hundreds of thousands of his codescendants, citizens of the so-called autonomous region of the Germans on the Volga, were given the order to leave their native homes in twenty-four hours. This was the first massive-scale deportation conducted during the World War II period.

The war brought hunger. I still remember the bitter taste of potato skin. But there were happier moments, too—such as eating an omelet made from egg powder, sent as food aid from America. My mother also displayed unusual entrepreneurial talent by finding a she-goat from a rural area. My family was living at that time in a small apartment building, which had a detached wooden porch. This structure played an important role during the first couple of years of the war—it was the goat shed. I was in charge of making sure that the goat had enough grass. Very often I would pick up my textbooks and accompany the goat to the garden of nearby Kazan University. For some time the goat was able to give two glasses of milk, one for each of my younger brothers.

I spent some of the wartime summers in a small town with my granny, my mother's mother.

Several times, Granny took me to the local market to sell some of her products, like cucumbers. I had to help her by being a little salesman. In many ways, it was a great embarrassment for me to tell the customers how many kopecks or rubles they would have to pay for a kilo of cucumbers or tomatoes. We kids of socialism never missed our share of brainwashing, and we hated free enterprise on ideological grounds.

Even with so much loving and caring, Granny was unable to influence my emerging class consciousness. I felt much happier sitting in the attic among the books left by my youngest uncle, Faruk, who at the age of nineteen—only ten years older than me—was taken into the army. For me it was like a real celebration during the time of the plague to live the lives of the heroes of classic Russian literature.

Granny assigned me a very important job: to help her with her letter writing to Faruk and my three older uncles, all of whom had gone to the front. My memory is still full of the unsophisticated stories she told her kids in her letters—stories of the latest developments in the neighborhood, on the streets, and in the family.

Faruk never came back. He gave all of his blood for the cause of victory. His older brother Zagir also did not come back. The two twins—the eldest, Khassan and Khusain—finally returned, humiliated, wounded, and handicapped for the rest of their lives.

Society developed a special procedure, a sequence of different politicized organizations, which evolved as we grew older. Until the age of eight, all of us were Octobrists, the most junior level of political maturity. In the third year of elementary school, we were elevated to the next level through a special process of Marxist-type baptism. This made us Pioneers—Leninists. Marching with drums and horns, we had to wear ties. The sacred ritual consisted of a standard exclamation: "The Pioneers, for the fight to defend the cause of Lenin and Stalin, be ready." We would loudly shout, "Always ready." Often boring, it was for us a game. We were kids.

In 1947, two years after the war was over, I was rewarded for my achievements and sent to Moscow with a group of Pioneer leaders. We were given the privilege of having "access to the body" of Lenin in the mausoleum. I was very disappointed. I discovered that the face of the founder of our country was becoming overly yellow. The immortality scientists, I thought, were not doing a good job.

At the age of fifteen, I was invited by the director of the school for a serious conversation. There was another boy in the office. One year older than me, he had just finished his term as leader of the school's Komsomol organization, the highest ranking group for young people, and he had to prepare himself in his final year of high school to go on to the university. The director tried to explain to me that the time had come for me to take the role of leader—as Komsomol secretary.

Too young to act consciously, to be spurred by serious assessment, I acted on instinct. When I was asked to take this position, I answered that I was an inappropriate choice for such a job and that I had to refuse. Somehow I was strong enough to resist the pressure that was applied on me during several talks. I felt a natural discomfort in playing the role of the boss. I was unable to accept this type of activity, most likely because it was compulsory by nature. These obligatory activities seemed contradictory to the interests that had begun to really take hold in me.

This was a real turning point in my life which made me different from what I would have been had I followed the established stereotype, the established tradition.

I felt that there must be a conflict between adopting the duty to issue commands and to organize people, and thinking freely about different problems in science. Perhaps that is the difference between organizing and analyzing. Anyway, I realized I had inherited this conflict from my father; it was something I could sense when I was in his library looking through his personal collection of books.

To my parents' great credit, no pressure was applied on me to reconsider when I finally told them I had informed the director that I would not take the position as Komsomol secretary. I felt such a joy at the prospect of absorbing myself in my own work. It was exciting to win chess games, to find mathematical solutions, and to tackle intellectual problems. This joy was exhilarating; it was overwhelming. I thought I should not betray it by being decorated with titles.

My family had always supported all of my efforts in academics, perhaps because my father may have felt unfulfilled in that dimension. From the beginning I was regarded as a successful student. I got good grades and I picked things up quickly. I was also very attentive in class.

However, at the age of twelve or thirteen I was forced to discover that I was unable to solve some of the problems given to me by my teacher. This had a very strong impact on me, so I decided to try to overcome these problems by myself, without asking for help.

First, I began to read some related material in my father's library, which somehow became an important trigger for me. I became intrigued and absorbed by the intellectual challenge set up by mathematics. Before long, I was regarded as someone who might become a mathematician like his father. I began to look for different ways to increase my knowledge of mathematics and its problems. When I discovered that Kazan University had a special evening course for school children, I enrolled.

During this period, a very important change took place in me. As I slowly moved from a fascination with geography to a passion for history, I discovered that I preferred the pure intellectual challenge set up by mathematics. Although I toyed with pursuing geography and history, I was concerned by their limitations: after all, the world had only two poles, and history repeats itself. But intellectual adventures in areas where pure mind can break new ground, like in mathematics, became more and more attractive to me.

At the same time, on the routine side of school life, I had always had the feeling that it was very boring to follow all of the instructions given in class, to do everything from A to Z. So I developed an approach: I decided to do what interested me first. I found that this resolution became increasingly important throughout my life, not only in school, but also at the university, where different compulsory political disciplines were drummed into us.

As I progressed through my elementary school and high school years, I fell in love with the game of chess. I began to devote a great deal of time to the game. In fact, I came back home later and later each day after chess parties and chess tournaments, all of which made my parents very angry. They considered this expenditure of time fruitless—a kind of art for art's sake.

For the bulk of my last two years in high school, I was sure that when I went to the university I would become a mathematician. At that time my encounters with physics and chemistry had been rather disappointing. Almost immediately I discovered that I had difficulty doing experimental work. Fortunately, the last course I took in physics began to change my orientation. More and more emphasis was placed on solving quantitative, analytical problems that required a bit of mathematical skill and some knowledge of physics. I became captivated by these types of problems. They combined the intellectual challenge of mathematics with the need for imagination in concrete terms. They were not confined by sixty-four checks on the chess board. Instead, these problems could be regarded as the precursors of theoretical physics.

As it happens sometimes, even a small external impetus can prompt an important internal revelation. The school was distributing application forms with a number of questions that we were required to answer. One of the questions, on social origin, I took back home to confer with my parents. They told me to put down the answer "from a white-collar family."

With great shock, I asked my father, "Does that mean we do not represent the dominant ruling class of our society, and we are not part of the dictatorship of the proletariat?" I was unable to accept that we were not a class but instead a social group that was considered auxiliary and subservient to the working class. In the company of my contemporaries, we usually described ourselves as "the proletarians of intellectual labor." "Proletarians of intellectual labor around the world, unite!" we would say.

As students of science, we simply were not told many things. Even among the professional physics community, for example, there was a campaign that suggested that certain elements of Albert Einstein's theory of relativity were doubtful. Other important contributions of Einstein and Niels Bohr were regarded with disdain, and some aspects of quantum mechanics, which were among the highest achievements of mankind during the twentieth century, were discredited. The views of Einstein, for example, were called "subjective" and "idealistic." Some critics said that the uncertainty principle in quantum mechanics was a contradiction to the Marxian theory of dialectical materialism.

We were told that nearly every invention was discovered in our country. Even many classical discoveries were attributed to Russian inventors. In every field we were given the names of the Russians who got there first. We were told that long before the Wright brothers, Russians were already in flight. And, of course, Russians invented the radio before Marconi developed the very same device, . . . and it went on and on. We were, however, sufficiently mature to enjoy the jokes that ridiculed such statements.

The inclination to take credit for everything was part of a larger movement under way in our country, referred to as "our struggle against cosmopolitans." Unbeknownst to those of school age, provincial ideological repressions were being waged in the large intellectual and cultural centers of the country. During the late forties and early fifties, the label "cosmopolitan" was given to anyone regarded as having insufficient appreciation for our national culture.

I experienced this phenomenon firsthand. Just before leaving high school, I had to take one final exam before getting my diploma. I was regarded as the obvious candidate for getting one of the school prizes, the Gold Medal. I had only to do well on this final set of exams.

We were asked to write about one of the early Russian intellectuals who had been prominent during the reign of Catherine the Great and, later, of her son, Paul I. This was Aleksandr Radishchev who, during the late eighteenth century, wrote an essay titled "The Journey from St. Petersburg to Moscow." In it he criticized the system of serfdom in Russia.

When our exam subject was announced, I got to work compiling everything I knew into a brief essay on Radishchev, complete with hard facts and a number of quotations. I stressed the importance of this writer during Catherine the Great's period, and I underscored the influence that the French Enlightenment had had on his thinking.

Upon completion, we waited for several days to receive the results. One evening my father came home and told me he had received a telephone call from the director of my high school, who said that he was calling about an embarrassing matter. He wanted to warn my father that a terrible thing had happened with my exam. "How was it possible," he asked my father, "that during the peak of our country's fight against the cosmopolitans, the son of an important official could show such respect for the influence of the French Enlightenment?"

"Our great writer," the director continued, "was not influenced by French cultural or political developments; his only inspiration for such ideas was the influence of Russian revolutionary reality."

My mistake cost me a rather low mark; and obviously, no Gold Medal was awarded. With a Silver in my pocket, my friends and I went to celebrate commencement night at a big masquerade ball for hundreds of graduates. At the entrance to the ball, we discovered that many of our contemporaries were dressed in costumes, and some had masks. In a small shop in the lobby, we bought the only thing that remained—two masks, one of a wolf, the other of a smiling clown. We put them on either side of one of our friend's head, adding a hastily written sticker, "The Two Faces of Wall Street." To our surprise the judges decided to award this mask first prize.

It was not too difficult to appeal to anticosmopolitan feelings in 1950.

4 / Moscow State University

➢

With high school behind me, I had to think seriously about what I would do in the future. Unlike most other students from my school, I was set on going to Moscow. I knew that I wanted to leave Kazan and its provincial intellectual life. However, I still oscillated between mathematics and physics. Even after my exams were over, I was still uncertain about what direction to take.

In such an indecisive state of mind, I was sent to the Russian Youth Chess Tournament to represent my city of Kazan as the local title holder. The event was held in Sverdlovsk, the capital of the Urals. I traveled in an old converted transport plane, and the passage was long and arduous. I arrived absolutely exhausted after many hours of travel.

I lost the first round of the tournament, but later I managed to get a grip on myself. Still, by the end of the tournament I had placed only third in my subgroup, half a point short of an invitation to the finals for the Russian Republic. In any case, the Federation Tournament would have been in direct conflict with my attempt to enter Moscow State University. In spite of bad luck with my chess, would destiny give me a better chance in a scientific career?

At the end of the tournament in Sverdlovsk, I spoke to my parents on the telephone. The deadline for applying to the university was only days away. I was still unable to decide to which department I should submit my application.

I approached my final decision as a gamble. Talking to my parents across a crackling line from Sverdlovsk, I asked them if it would be too late to have my documents sent from the mathematics department to the physics department. When I returned, I discovered that my papers had indeed been sent to the physics department.

Within a couple of days of my return, I was invited to go to the university for the entrance interview. I felt eager to go back to Moscow, this time by myself. I had not been in the city since we moved to Kazan, but despite all these years I could still remember some of the streets and some of the buildings around the dormitory where I had lived with my parents.

It was intriguing to visit the same dormitory that had been part of my life during the earliest years. This particular dormitory had been reserved for postgraduate students, which my parents had been while they were at the university. Given the chronic housing shortages in postwar Moscow, many former postgraduate students who started out in this dormitory or others like it were still living there years after they had moved onto professorial jobs at the university and other places.

I remember distinctly the exam I took during that week in Moscow. I was asked a number of questions about physics, none of which I found particularly difficult. What I found most curious and disconcerting of all, however, was the final question. My examiner asked if my parents would be able to provide support for my living accommodations in Moscow. The examiner asked me if they could rent me an "angle," the Russian slang for the corner of a room in a communal apartment. I was confused and embarrassed by the question. I didn't know it, but people on pensions, who needed additional money, rented out corners of their one-room habitat.

The notion of having a stranger rent me a corner of his one-room dwelling seemed completely foreign to me. But what was I to do? If admitted, I would be given the special opportunity of having a Moscow stamp in my passport.[1] I had the distinct impression that my answer to this question would determine my chances of getting into the university.

When they asked how much money my parents would be willing to spend to find me a corner, I hesitated. I had no idea what the monthly cost of a corner would be, so I made a simple guess: "Twenty-five rubles." Suddenly there was a burst of laughter. Twenty-five rubles, at that time before monetary reform, would today be equivalent to a few dollars in America.

I knew as soon as the words left my mouth that I had said something ridiculously stupid. I can only think that my examiners were so amused by the naïveté of this provincial standing before them that they decided I deserved a bed in the student dormitory. With that, I was given a docu-

1. A stamp in a citizen's "internal" passport was and is required. It is essentially permission to live in the city.

ment that entitled me to a bed in one of the oldest dormitories of Moscow State University.

Stromynka dormitory, as it was called, was a vast, quadrangular building with a poorly kept interior courtyard. It boasted no less than 10,000 students, with each academic department occupying separate floors. The walls of our castle (or monastery) were on the banks of the Yauza River, a tributary of the Moscow River. According to legend, our Stromynka dormitory was originally built as military barracks before the time of Peter the Great.

It was not without sacred trepidation that I entered for the first time that temple of Russian science, Moscow State University. In Stromynka dormitory I was assigned a corner of a room on the fourth floor. Occupied by ten students, it was the biggest room around. There was only one long desk in the room, which was used mainly for meals. Most of our studying was done in the sizable library not far away. To study in the library, however, one would have to get up very early to be there in time to get one of the precious few seats.

All of these problems, however, were completely overshadowed and overwhelmed by the true atmosphere of camaraderie among roommates and classmates. We came together from different places in the country. However, one of the distinguishing features of life at the university in the late forties and early fifties was the influence wielded by a significant number of students who were veterans of World War II. They were older and more seasoned, and all of us younger students treated them with a little awe and a great deal of respect. In exchange for the wisdom the older generation of students brought to our community—with a feeling of generous indulgence toward us "green adolescent boys"— we helped them get through cumbersome and difficult mathematical formulas and theories, which were not easy for many of these men. Of course it was tough for them to compete with the scientific wunderkinder, whose student years had not been interrupted by the war.

Somewhere inside this giant student anthill there was another freshman, whose name was known only to very few friends and teachers at the Department of Law. His name was Mikhail Gorbachev. I never had the chance to make his acquaintance while at the university. Perhaps the students of physics were too isolated from the students of social sciences.

Later in an interview, Gorbachev confessed that before coming to the university from the southern rural area around Stavropol he had planned on becoming a physicist. Perhaps if his plans and provincial dreams had materialized, we would have met in the same classroom. But he was unable to get into physics. The most plausible explanation for this would be that the requirements for physics were extremely severe. Applicants had to prove they were entitled to security clearance. Perhaps the most serious guilt carried by Gorbachev, and quite a few other people of my generation, was that they had lived in territories occupied by the German army during the war. It was a relevant factor, even though it had not been their own conscious decision. After all, they were only young kids and, like Gorbachev, were not even ten years old when the war broke out.

Physics at Moscow State University had an old tradition from the very moment of its establishment. Mikhail Lomonosov, one of the greats in the history of Russian science, who participated in drafting the charter of the university in 1755, insisted that the chair of physics be instituted within the Department of Philosophy.

In the nineteenth century the physics department at the university had won international recognition, especially through the pioneering work of the famous Russian physicist Peter Lebedev, who, at the end of the nineteenth century, was able to perform one of the most elegant and masterful experiments of the age: he measured the effects of light pressure. Though the phenomenon had been predicted long before, it was considered one of the principles of the Faraday-Maxwell theory of electromagnetic fields. Its experimental proof required tremendous ingenuity. He proved the theory and the validity of the hypothesis of Johannes Kepler, who, in 1619, concluded that the peculiar shape of a comet's tail could be attributed only to the existence of solar light pressure.

In its own way, Lebedev's experiment created the bridge to modern-day physics which is based on the intrinsic relationship between energy and mass. His experiment paved the way for Einstein's greatest discovery, which changed the way of life and the way of thinking of mankind in the twentieth century, an epoch dominated by nuclear energy. Lebedev and quite a few other prominent figures in the history of Russian science at Moscow State University were seen everywhere around campus on canvas paintings and in sculpture. The buildings, the labs, and the lecture halls of old Moscow State University, in the very heart of Moscow, bore the names of these great ancestors.

I remembered these giant names from the stories my father told me of his student years in Moscow. We students of the early fifties were eager to meet the savants of our time—the keepers of the flame of the old traditions of science, the torchbearers of knowledge. But when I arrived at the university in 1950, to my big disappointment these professors were no longer on the list of faculty. The university had become quite a different place from what it had been during my father's student and postgraduate years. Many of the names of the brightest physicists were missing. Within a few years after World War II, the best physicists were dropped from the list of professors.

Among those who were forced to leave was Igor Tamm, one of the most popular young professors of my father's class. A textbook of his had played an important role in my early fascination with physics. During my father's time, Tamm had been a young professor who had been actively engaged in the life of the university. Years later, even though we used his textbooks, we all knew that he had been expelled and his name blackened among the new breed of faculty.

The faculty had been taken over by a group of second-rate scientists—mediocrities who used ideological warfare to drive out the likes of Tamm. They accused him and others of his caliber of using an "idealistic" interpretation of modern physics without giving enough attention to Russian national physics. With these charges, little by little the best professors were eliminated. At the same time, instead of getting a first-hand approach from creative scientists, the younger generation learned that science was a subject of established, given facts. This was not the way to absorb the flavor of the creative process of science.

We students got the sense that there was something fundamentally wrong with the Department of Physics. While we were eager for our courses in physics and the accompanying lectures on mathematics, chemistry, and engineering, we came to despise the lectures that distracted us from our main passion in life. These lectures materialized in the form of compulsory classes on ideological and military topics. A sort of competition arose between the ideological indoctrination lectures on the history of the party and Marxist philosophy and those on the military.

Despite my boredom, I was far from trying to ignore these lectures because I knew how dangerous it could be for a student were he to do poorly in such classes. Students who did not pass courses on the military were in danger of being drafted into the army. Without passing these courses, they would not get the diploma of a reserve officer.

Our ideological obligations were another matter. In addition to the history of the Communist party, every student received courses in Marxist social sciences and philosophy. They were separate from our science studies, and they were absolutely compulsory. We had to be ready for very strict examinations twice a year. During the first couple of years, we were expected to learn the important facts about the history of the party.

The biggest impression left on the students of my generation, after learning this history, was that the most important priority throughout the history of the party was the internal fight. The principal development in party life was to identify potential adversaries, enemies, and so-called deviationists—left-wing deviationists or right-wing deviationists: opportunists and social traitors, Trotskyists and Bukharinists. One could only "admire" the titanic job performed by Stalin in trying to maintain the purity of the party ranks!

Our next course was also not exactly what one could call science. It was Stalin's interpretation of the Marxist philosophy of political economy. We had to memorize lengthy quotations from Lenin and Stalin. Even those of us who could pretend to have an infinite memory would have to do the mandatory job of compressing the content of their voluminous works and reproducing it in handwritten notes to our teachers.

It was very time consuming. Even compared to the mediocrities who taught physics, the ideology teachers were laughable. During the lectures, the teachers had a great deal of trouble keeping discipline. It seemed that the students were always on the verge of bursting into laughter—a breakdown simmered just below the surface. The whole country was being educated in quasi-science, and so it is probably not surprising that the people who were teaching it were often intellectually and culturally underdeveloped. But then, unlucky party functionaries were converted into these types of teachers and professors.

All of this seemed utterly illogical to me and my friends. In fact, we regarded ourselves as superior in many ways to the Marxist philosophers who promoted these beliefs. This notion of superiority began to filter into our student folklore. We had one particular joke about a store that sold brains:

"How expensive is a kilo of physicists' brains?"

The answer: "Ten rubles."

"And a kilo of chemists' brains?"

"Thirty rubles."

"What about the brains of philosophers?"

"It would cost a hundred rubles per kilo."

"Why so much?"

"Imagine how many philosophers we would have to kill to get a kilo of brains!"

Imagine such a joke in Stalin's lifetime—and Stalin regarded himself as the greatest philosopher of all time! Many of the students did in fact believe that Stalin was a great scientist. The very language of his articles seemed impressive. They were punctuated with an Oriental style, full of simple wisdom. There were many legends among the people about how competent Stalin was. In every textbook or encyclopedia of that time, Stalin was given the title Father of every field of knowledge, every activity in the country, such as the Father of the National Car-Producing Industry and so on. It was only later that I heard one plausible explanation for the origin of his seemingly competent remarks. The story says that before Stalin met with professional groups, the KGB conducted special research, interviewing many people from the industry or work project to gain a perspective on certain problems from the inside. They would then brief Stalin with short, fabricated questions and remarks. That was how Stalin made himself a legend.

Within the framework of our ideological education, the Komsomol played a significant role. Everything in the country was run by the party; even family conflicts and personal problems became the business of the party apparatus. Often the Komsomol or the Communist party itself would be asked to intervene and play the role of judge. In the case of students, it was the Komsomol organization that provided the oversight for our intellectual and social behavior.

Once I had a scrape with the Komsomol organization. Only later did I realize how close I had come to getting into very serious trouble. My second year at the university, I was approached by the editor of a small newspaper within the Department of Physics, just prior to the fifth of December and the celebration of the great day of Stalin's constitution. The paper, which was quite widely read—in large measure because its pages were hung on the walls of the university—featured the lives of particular students on campus. The editor suggested that I write a special article for the celebration. He told me that the article should say "how happy you native minorities are and how grateful you Volga Tatars especially are to Stalin for his great constitution that has brought equal rights."

Without much thought and without any special motive—I was really quite busy with my studies—I simply rejected the request with what was perhaps an inappropriate aside: "Why so much fuss about it? It's very trivial." Only later did I understand at what risk I had put myself. In retrospect it is easy to see how my refusal to write the article could have been misinterpreted. In fact, there were some hints of an attempt to use my unwillingness on that occasion as evidence of dissidence.

Later I heard from a friend of mine that he accidentally discovered a note written by a fellow student that had been lying on a table in the library. The note, which painted me as a "dangerous element," said that the time was coming to show everyone who "he [Sagdeev] really is, and what his political face really looks like." During this epoch of Soviet power, an incautious anecdote, a joke, an expression, or an exclamation could lead to big trouble. All the same, we couldn't stop joking. That was probably our most effective medicine for mental survival.

I remember the introduction of a new discipline at the university and everywhere in the country known as civil defense. Just as Marxist philosophy had been conceived to make us immune to the hostile bourgeois ideology, civil defense was conceived to make us immune to nuclear attack. We students knew very well that both concepts were garbage. Unfortunately, we were unable to completely reject them or deter them. We were wise enough to look for a realistic compromise, a kind of minimal deterrence.

As we were to discover over the years, the common view of Marxism, as a very rigid testament, was completely wrong. Quite the opposite. Marxist philosophy was an extremely flexible instrument in the hands of true demagogues. For example, an unbiased analyst could hardly accept that the denial of classical genetics, based on the existence of genes and chromosomes, served the interests of the proletariat. However, it was the evil genius of Trofym Lysenko who used this argument to dampen and, finally, completely annihilate his opponents and fellow biologists. He was capable of mobilizing the army of Marxist philosophers to prove that classical genetics is alien to dialectical materialism. However, at the most dramatically critical moment in 1948, when he felt especially endangered, he asked for the blessing of Philosopher Number One, an endorsement that yielded victory to Lysenko. The result was the decisive elimination of the true science of biology and genetics from the face of Soviet science for many years to come.

Physics was one of the most fortunate scientific disciplines. Compared to virologists, physicists did not suffer on any scale as a result of ideological considerations. However, even if the success of demagogues in physics were a rarity, it was precisely the case with Moscow State University that those with the best brains in physics were kicked out of the department by the time my contemporaries and I came to the student bench.

5 / Only Physics Makes Sense

The freedom-loving student community counteracted indoctrination and brainwashing through a growing self-consciousness as physicists. This was reflected in the bravado songs we sang at parties: "Only physics makes sense. Everything else is nonsense." And: "Philosophers and philologists are blockheads."

We were more and more involved in the life of physics. If I had been asked at that time how one defines "physicist thought," my answer would have been that physics approaches everything from basic principles to establish a basic identity and basic features, to ask the most appropriate direct questions, to find the components of the overall picture, and to get rid of the components that are not so relevant. Only then can it build a simplified model that reproduces the basic behavior of the object under study.

My academic achievements were shown appreciation in the form of a fellowship given to me during my third year at the university. I received a fellowship established to commemorate Nikolai Morozov, a nineteenth-century revolutionary. A dissident sent by the government to the Shlisselburg Fortress, one of the traditional places for such punishments, Morozov became famous for his endurance, surviving more than twenty years in an isolated cell. He was regarded as one of the most influential political and intellectual figures of prerevolutionary Russia.

Among my friends, political activism was not considered a desirable use of our time. Many of us were more and more obsessed with science, and it would have been a sheer waste to spend time on such things. But there were quite a few students who especially entertained duties in Komsomol or later in the party. They knew they would eventually be rewarded with better opportunities for fellowships, postgraduate appointments, and future career choices.

Perhaps one of my most intriguing roommates during my first year in the university dorm was a soldier who had come to Moscow from Kamchatka, one of the furthest outreaches of the Soviet Union. His name was Oleg Lavrentiev. We never knew if he had seen action. In any case, he had the air of maturity that you would expect from someone five years older. But more than maturity was the aura of mystery Lavrentiev carried with him. We sensed that there was something different about him. He seemed smart, but we sensed that he had some kind of additional authority. His notebooks and the way he talked to us gave us the impression that he knew something we didn't.

One evening when I returned to our room, everyone was in a state of excitement. "We lost our Oleg," my roommates shouted. "Several people in black raincoats and black hats came and took him and all of his belongings!"

I was absolutely terrified. "You mean to prison?" I asked. No one knew.

The next day I saw Lavrentiev in the lecture hall. He appeared to be in a very good mood and in fact looked even more confident than usual. Indeed, he gave off an air of real importance. Rumor was that he had been given a special apartment in Moscow. The people who came to pick him up in the black limousine were most likely his sponsors.

After that, we noticed that Lavrentiev was given the unique chance of having access to the best of everything the university had to offer. He even had lectures conducted for him alone. All of us thought that he must be some sort of genius working on something top secret. It was not until I finished university and began working at the Kurchatov Institute of Atomic Energy that I discovered his secret.

I believe his is one of the most dramatic stories of the scientific revolution.

The first nuclear bomb was made as a result of a fission reaction. Theoretical physicists had known since the late twenties that the process that moved in the opposite direction, fusion, would produce an even greater amount of energy. But to materialize a fusion reaction would require that nature overcome the electric repulsion between nuclei. The most natural condition for overcoming electric repulsion exists within a high-temperature environment. The temperature has to be so high that at first it was almost unimaginable that fusion could be achieved in a man-made environment. Physicists thought the energy of the sun, and of the stars in general, could easily be explained on the basis of such thermonuclear reactions.

While physicists in such top secret places as Los Alamos National Laboratory and, by that time, secret Soviet institutes were contemplating how to overcome this difficulty, they did so in order to create an even more powerful destructive weapon: the hydrogen bomb. Very few people then tried even to think about how to convert the thermonuclear fusion process into a different channel for peaceful uses. Experts realized that such an outstanding problem would require at a minimum a Los Alamos–type approach. Driven by sheer curiosity, some of the researchers were brave enough to come up with specific guiding ideas. It was suggested that by generating strong acoustic or sonic waves, it would be possible to heat matter to the necessary temperature. Even a student could understand how naive it would be to reach critical conditions for a fusion reaction in such a way.

The experts knew then that even the best ideas on heating matter to critical thermonuclear temperatures would not resolve the even more difficult issue of keeping the substance at such a temperature long enough to enjoy the release of energy from the reaction.

The young soldier Lavrentiev, while on the distant Kamchatka peninsula in the army, was not sleeping nights but trying to devise what would now be called the electric confinement of hot matter or plasma. He got this idea while he was still serving duty. Like many hundreds of thousands of other people who had something extremely important to communicate to Comrade Stalin, he sent a letter. It was remarkable that his letter wasn't thrown out or lost.

One day in early 1950 Lavrentiev was called in by his military bosses and was told that they had instructions to send Lavrentiev to Moscow State University for a crash course in physics. Very few knew that the young soldier Lavrentiev had fallen under the personal patronage of Lavrenti Beria, Stalin's KGB chief and the government's coordinator of the still-infant nuclear program.

I thought Lavrentiev's status was curious at the time. I can still recall being startled one day after an informal chess game between us. Lavrentiev asked, "Do you know the name Sakharov?"

Andrei Sakharov at this time was deeply embroiled in the H-bomb project. The scientific proposal contained in Lavrentiev's letter to Stalin, however, had been passed on to Sakharov for an evaluation. While its potential application did not have any direct bearing on the bomb project, the proposal had obvious relevance to the controlled fusion program for energy production.

In many ways, I felt sorry for Lavrentiev then, and still do. The idea of making him into a blitz scientist ultimately failed, of course. Physics is a science that consumes a lot of time. After the first year of making some steps forward, his lack of fundamental knowledge eventually set him back. Years later I got a perspective from a different angle, from Sakharov himself, on what had happened to Lavrentiev. Apparently the letter that Lavrentiev had sent to Stalin was delivered to Sakharov for his assessment. Sakharov shared it with his closest colleague and teacher, Igor Tamm. At the conclusion of their evaluation, their rather straightforward calculations convinced them that the electric field would not work in confining hot plasma and isolating it from the walls of the container. Even a strong electric field would be unable to confine any noticeable amount of hot matter due to one of the most important properties of hot, ionized matter—the plasma itself. The positively charged nuclei and negatively charged electrons would produce a net zero charge which would nullify the effect of any applied electric field. What Lavrentiev did accomplish, however, was to trigger their thinking. Soon Sakharov and Tamm came up with the revolutionary idea of using a magnetic field instead of an electric one. Lavrentiev had thus led them in the direction of finding a plausible method of confinement.

Sakharov and Tamm were not only great scientists; they were extremely generous and decent people. In a way, they had a legitimate right to qualify the idea in Lavrentiev's letter as naive and uninteresting, and then to proceed with their own invention. But instead they said that even though this idea would not work for fusion directly, it had been an important stimulus for them to find the right approach. With that in mind, they suggested that Lavrentiev be brought from Kamchatka and given the chance for a proper education, presumably at Moscow State University.

Lavrentiev was a sad character to me. He came to Moscow under the infamous patronage of Beria, thus promising very little sympathy from the scientific community. The irony was imprinted in his very name. *Lavrentiev* literally means, in Russian, "son of Lavrenti." The grand educational experiment, the blitz lectures especially developed for Lavrentiev by the best professors of Moscow State University, did not work. Instead of being boosted with a giant leap forward in knowledge, Lavrentiev lost momentum and eventually came back to resume normal classes with the rest of us.

But in some way, I believe, he lost his stamina. For the next decade he persisted in trying to prove his original concept for electric confinement. Of course, he was given the chance to collaborate and even to do some simple experiments, but the tragedy was that he unilaterally refused to accept science as a product of collective work. Eventually, he was asked to leave the Kurchatov Institute and Moscow, and he was subsequently sent to one of the big installations in Kharkov.[1]

Without a doubt, the most important event during my years at university was the death of Stalin in March 1953. All of us were taken by surprise. Most of the Soviet public was not ready to accept the fact that Stalin was as mortal as everyone else. We were in our dorm room when the news came to us over the radio. The students were at first stunned into silence. The following two or three days were among the most confused and chaotic that I can remember. They were an odd mixture of embarrassment, turmoil, excitement, and despair. What would happen to our poor country—to poor us? We were lost.

We did not yet know that shortly before his death, Stalin, addressing his closest associates in the Politburo, had given them his sacramental sentence: "Poor kittens, without me the imperialists will strangle all of you very soon."

These days were the first of deep discussion, too. We students suddenly discovered an interest and a need for an internal political life. Without consciously realizing it, we started the first political brainstorming of our lives. We debated the accomplishments of our nation under the leadership of the great Stalin and even made some reasonable assessments of what might happen to the country: who would succeed Stalin and what kind of problems we were going to face. We never tired of talking to each other about deep political problems, because we had never had the need to talk in such a way before.

1. The tragedy of Lavrentiev's unmaterialized intellectual potential came from the attachment Beria felt for him in the early fifties. Lavrentiev's colleagues at the Kurchatov Institute never forgave him for the moments he dialed Beria's office to talk to his patron. Sakharov, however, was above all of these prejudices. He found warm words for Lavrentiev in his memoirs. When Sakharov died, in December 1989, I received a letter from Lavrentiev. In it he told me of his last conversation with Sakharov, when he was invited to Sakharov's apartment in Moscow in November of that year. Lavrentiev was very touched with the way Sakharov spoke to him at the tea party when Lavrentiev asked Sakharov's advice about his current research—which probably is, in a strange way, a continuation of his early attempts.

We guessed that Georgy Malenkov, one of the youngest members of the Politburo, would succeed. It was he who had recently delivered the most important speech at the big, solemn celebration, and at the demonstrations and military parades, Malenkov had stood closest to Stalin on the mausoleum. We also guessed that it would not be a very stable leadership and predicted that there would be a number of personalities vying for power, creating a real power struggle. After all, we were not bad students of the party's history.

In fact, that is virtually what happened. Malenkov was indeed appointed first secretary of the Communist party. And his rule was indeed unstable and rather brief in duration. He barely had the chance to address the nation, to say that he was going to place greater importance on improving living conditions and on increasing the production of consumer goods. Without much warning, he seemed to disappear from the horizon, replaced by Nikita Khrushchev. It was Khrushchev who was most instrumental in finding a way to eliminate Beria. We could only follow the official statements, but we sensed how dramatically things were developing.

Another very important issue of the time was that of political prisoners, many of whom were released after Stalin's death. Their release had a very strong impact. It was the first silent confession that something had gone wrong under the great leader, the father of the people and the genius of humankind. Naturally, we did not realize it at the time, but this was only a very small beginning. It wasn't until the Twentieth Party Congress in 1956 and Khrushchev's speech in which he attacked Stalin and the "cult of personality" that we understood, with surprise and shock, the extent of the wrong that had been done.

On our return to Moscow the next fall, instead of going back to the old dormitory and lecture halls, we settled into a new building on a new campus. It was a new start, and it seemed to go hand in hand with our new birth of self-consciousness.

One of the most memorable events of my time at Moscow State University occurred during the first autumn of the post-Stalin era not long after we students had moved into the new dormitory. It was the first time I met Lev Landau. One of my roommates, an acquaintance of Landau, suggested that I be one of the hosts for Landau's forthcoming visit. I was very proud to be asked, but I was also extremely nervous before the meeting. Though Landau was only forty-five at that time, he had become a living legend among young students of physics. We knew of course that he was the pupil of the great Bohr, having spent a few years in the sacrosanct brain quarter of contemporary physics, the Copenhagen school.

Of the many scientific achievements of Landau, the most important was his theoretical explanation of the phenomenon of the superfluidity of liquid helium. Landau was already very famous for his absolutely unchallenged approach to introducing the topic of theoretical physics to younger students. Despite the physics department's obviously chilling attitude toward Landau, many of us had already acquired several volumes of theoretical physics textbooks that Landau had cowritten with his longtime colleague Yevgeny Lifshitz. I knew this was going to be my first meeting with a living scientific genius.

I will never forget my first impression of Landau. He was a tall, slim man, with apparent ease of carriage. He had feminine hands and a rather pleasant voice that sometimes created the feeling that he swallowed the endings of words and sentences. His friendliness immediately disarmed us, and the general atmosphere of our first meeting became free and natural. He clearly took an interest in the wonders of the very rich campus and the central tower of the university, the last gift of Uncle Joe to Soviet science. He was in a jocular mood, commenting on how easily we might be spoiled in such an atmosphere of magnificence. After our sightseeing, we went to the small room of one of my classmates and enjoyed a student-style tea party.

I remember Landau asking us what we planned to do next. When it was my turn, in a trembling voice I said, "I dream of being a theoretical physicist."

Landau turned to me and said, "So what can you do? What have you learned at university? Can you handle a standard set of mathematical problems? Do you feel you would be able to do it spontaneously, with ease, without even thinking?" Then he paused and looked at us.

My reaction to Landau was disappointment. Instead of talking about the outstanding problems of physics, he spoke about something that had to be developed as a routine mathematical procedure, without even the need for thinking. Catching our surprise, he went on.

"This is very important," he emphasized. "It is vital to do such things automatically without thinking, because you have to keep your brains free for creative thinking about physics. Mathematics should be a language to you, one that you can speak fluently."

Then one of us said, "But why should we spend so much time developing such a skill when we can take a textbook of mathematics and find there any formula we would need?"

Landau answered with an ironic smile, "I hope you are not planning to become an accountant."

We were completely deflated by this remark. Then he added in a casual way, "When you have completed your preparations, you are welcome to come to me. If you pass my examination, you can become a member of my group. But there are special rules and you have to prepare yourself."

So that was the result of our first meeting. The whole afternoon left me in a spell, so strong was the motivation he inspired in us to pass his exam. The Landau minimum examination was by that time world renowned. As we had heard, it was comprised of about half a dozen different topics.

The day after our meeting with Landau, a friend of mine—a Muscovite—and I joined forces to study for the Landau minimum. We designed a schedule for ourselves to study and develop the mathematical capability to meet that arduous set of requirements necessary to pass his examination. This had to be done outside the conventional framework of university studies. My small room in the dormitory was converted into the headquarters and think tank for such preparations. We studied nearly around the clock for almost three months in this fashion.

At last, we decided that we had substantially refined our computational mathematical skills and were ready to sail on this unexplored sea without accountant's books and take the exam that would bring us to the new world of theoretical physics.

Coming out of our self-imposed isolation, we were completely unaware of what had been happening on campus. On the completion of our studying binge, we discovered that something was different in the department's atmosphere. Our friends were talking about the forthcoming conference of the Komsomol members of our department. So what, we thought. This is a very routine and a very dull type of procedure.

But this time, something else was in the air. Something was beginning to stir. Why this new awakening should come first from the physicists' community was not clear. Somehow we felt intellectually freer. I have often wondered about our sense of freedom, and I have concluded that perhaps we felt as though we were subject to a different set of rules, the ones that describe the order and supreme harmony of the universe in nature.

The Komsomol was attended by nearly a thousand students. It started in a rather normal way, with slow recitations by the principal speaker in the presence of the leading faculty and the party bosses on the podium. Suddenly some kind of impulse came from a few of the older students. It was a spontaneous eruption of protest. All the feelings we tried to hide, which had been communicated privately, suddenly burst to the surface. Eloquent speeches were delivered from the floor about the state of learning in the physics department. While we did not go beyond the domain of the department, the message was loud and clear. The Department of Physics and its intellectual atmosphere had to be changed.

In the course of the next few hours, one could sense that the temperature was rising and that sooner or later the pent-up feelings would boil over. It was the first time in all of our lives that we had experienced such a thing. People were talking about issues that we had even been afraid to name. Suddenly it burst into open discussion.

I can remember feeling my neighbor's elbow gently nudging my side in reaction to the defensive rebuttal being offered by one embarrassed and unsettled professor. So I was not alone! The chairman of the department, a tall, heavy professor of theoretical physics, Professor Sokolov, was asked directly why such people as Landau were not teaching here at the university. "What is preventing the university from inviting the best Soviet physicists?" The chairman answered that Landau rarely quoted Lomonosov in his works. That immediately created an explosion, a burst of laughter from the audience. Students continued to complain that the country's greatest scientists were no longer a part of university life and that again they had to be brought to the department to replace the mediocre professors who had taken their positions.

We understood that many of the most brilliant scientists, like Landau, Tamm, and Lev Artsimovich, were also engaged in national security programs, but we demanded that they be brought to the campus for lectures. Surprised and humiliated, the professors of the department tried to tranquilize us. But what they said to us seemed ridiculous, and its only effect was to bring to a boil even stronger protests.

The four-hour meeting ended with a set of specific demands: bring back the best brains to our lecture rooms and fire the most odious and mediocre teachers, who were spoiling our appreciation and understanding of physics with attempts to indoctrinate us with unnecessary philosophical banalities that were alien to creative science.

The conference was over, but for several days afterward we still felt tension in the air. Somehow the exhilaration of the protest seemed deflated. We were anxious, but not particularly surprised, when one by one we were called into the party committee for a review of the incident and our roles in it.

The party secretary who was investigating the protest was an older man. It was obvious that he had been a veteran of the war. He carried with him the traces of his former military uniform. In a rather stern voice, he methodically asked me question after question. I was startled, however, by the direction his probe was taking. He posed the question "Did you see Landau at the university before the Komsomol conference?"

I was stunned. Although my friends and I had never made any secret of Landau's presence on campus, I was genuinely surprised that the visit had come to the attention of the higher circles. I replied, "Of course, yes. He was here." And then it was immediately clear to me that the party secretary was trying to make a case that it was Landau who had served as the agitator, provoking the students' protests and demands. He was trying to draw the direct and logical line that all of the events, and our participation in them, were linked.

"We only spoke about science," I told the party secretary. "It was just a sightseeing visit to the campus."

But the tone of the discussion took a decidedly intimidating turn. It was clear that they were going to piece everything together to come up with a case against us.

Other people were called in over the next few days. The whole investigative process took two weeks. Throughout this time, an air of nervous expectation hung over the entire department. At the same time, the statement we had adopted at the end of the conference was making its way to even higher circles outside the university complex. We were far from these circles, so we could only guess what the final outcome would be. We all believed that the final verdict would come directly from the Central Committee, and when it did come, some weeks later, all of us were left speechless.

We never saw the statement. It was explained at a meeting hastily called by the university officials. It said that the introduction of important changes was imminent. The response was positive: our struggle had been considered just. Significant faculty changes were going to be made. From our point of view, the most odious figures were indeed removed. At the same time, the physics department got a new chairman, one of the close assistants to Igor Kurchatov, that indisputable leader of the Soviet nuclear program. Later we understood that the final consideration of the Moscow State University student revolt probably had not been settled without Kurchatov's participation.

At the same time, several new part-time chairs were set up in atomic physics. They would be held by Artsimovich, another close associate of Kurchatov who was already famous as an experimental physicist; Tamm, who was reinstated as a part-time professor; and even Landau, who was invited to give a series of lectures on selected problems of theoretical physics. This was our first taste of victory and exposure to real perestroika in our time. So it was with a pleasant feeling of victory on the political front that my friend Aleksandr Vedenov and I decided we were ready to take the first of Landau's examinations.

In early January 1954 I boldly picked up the telephone and nervously dialed Landau's phone number. "Ah, Comrade Sagdeev," he said. "Please come."

My teammate Vedenov and I were given separate appointments for exams in Landau's apartment. When my turn came, I made my way to the Institute of Physical Problems, the sacrosanct Soviet institute established by Peter Kapitsa in the mid-thirties on the outskirts of Moscow. This is where Landau worked and lived.

Landau's house occupied two floors. The rooms were small and cramped and, as I remember, somewhat dark. When I arrived, Landau opened the door and invited me to the second floor. He brought me to his room, probably his study, which was sparsely furnished. Besides a couple of bookshelves there was a couch in the center of the room. Evidently it was his favorite place, where he would do "science on the sofa." There was also a very small table that guests could use to solve their physics problems. Everything was quite simple and democratic, without any unnecessary formality.

His "Good morning, let's go to work" was businesslike but very friendly and inviting. On a small table he put a piece of paper with an expression of an integral for me to solve—the first mathematical routine Landau wanted me to calculate. I started to do the job after he left the room. I don't know how much time passed, but when he came back I had already completed my job. Landau looked through my work and determined that the final answer was right, but he commented on a funny omission. A couple of times I had forgotten to conclude the expressions under the integral with the always-to-be-present standard symbol *dX*. He reacted with humor: "The integral without the dX is like a man without his trousers." And then he wrote the second exercise.

In the course of an undefined period of time, Landau came and went. Altogether the exam took several hours. Everything was very efficient and very friendly. At one point he sat down on the sofa, next to the chair, and said to me, "My apartment is always open to anyone who is capable of passing the exam. I am not interested in students' political records or their ethnic background. I don't care if they have a criminal record. I am only interested in their abilities." Later I came to the conclusion that by so describing his approach to recruiting young scientists, he stressed the ridiculousness of the culture that existed in society.

I passed the first exam. But it was only the beginning of the long road I had to travel. I had to complete the entire "minimum," consisting of nine different exams. Landau told me that those who passed all the examinations would be entitled to his help, his interest, and his care.

Years later, after Landau's death, Lifshitz gave me a small souvenir of Landau. Landau had started this program of exams almost from the moment he finished his work with Bohr. In twenty-five years, forty-one students had successfully passed the Landau minimum. As I was one of the successful examinees, Lifshitz thought that I would like to have a souvenir of that effort: a photocopy of the written record Landau had kept. Next to the names of the examinees, there were marks by Landau indicating what they had achieved at the time the note was written. He must have attached special importance to the correlation of his successful students with their future accomplishments. Almost without exception, his pupils became prominent Soviet theoretical physicists.

6 / Sakharov's Installation

➢

While I was taking the Landau exams over the course of a year and a half, we students had to decide about our immediate future. Everyone was looking for the most interesting assignment for diploma work. We were, of course, excited about the developments taking place in the nuclear sciences. It was clear that the most interesting jobs could be found in this particular sphere. However, there were almost no open publications with job descriptions of this type. We lived in an era of utmost secrecy and suspicion. The only notices we could find came from the student grapevine, which passed from one generation of students to another.

Most of my friends were looking for diploma internships, which would open career opportunities that would allow them to stay in the same place. I felt different. I was rather confident that by the time I finished university, my Landau minimum examination would be over and everything would be taken care of by Landau himself. Of course, I hoped to be invited as a postgraduate to the famous Institute of Physical Problems.

Physics in the USSR at that time was gravitating to two different poles. One of them, the smaller, was the Academy of Sciences, dealing mainly with basic sciences and research—with a rather indirect involvement in sensitive applied programs. But such jobs were not given top priority. Everyone at the time was more excited by the prospect of going to the heart of atomic physics: the nuclear centers. Such installations were known as "mailboxes." They were so highly classified that they were not given intelligible names that had anything to do with the real substance of their work.

Entering this kind of physics had an enormous number of advantages. We knew, for instance, that the system in charge of such militarized sciences was the most powerful and rich in the country. In addition to higher salaries, employees would also benefit from special food stores and medical treatment. Furthermore, such institutions were surrounded by an aura, an atmosphere of enhanced respect in the country. There was a general feeling of pride in working for the military-industrial complex, though the very notion of such a complex came to us only much later with Eisenhower's famous farewell address.

We had praise for people doing such jobs. Pacifist sentiments were extremely rare at that time, and where they existed they were not spoken. This is perhaps because a strong feeling of vulnerability still existed after the war. After all, from early childhood we had had it ingrained in us that we were facing the greatest confrontation of all: that created by two rival ideological systems. If socialism were to be built in a world where its premise was challenged, then the nation would have to be vigilant. I think the very statement of this problem certainly gave purpose to party theoreticians and practitioners. But it also kept the citizenry in a permanent state of alert. We were born with the feeling that, as a nation, we were always in a kind of "circle defense." It developed into a siege mentality.

I think it would be fair to say that the basic symbiosis between the military and science was established long before World War II. In our history, we had been taught to admire and praise military commanders. The bright young commanders of the Red Army, like Mikhail Tukhachevsky, were applauded as being men of vision, farsighted thinkers who understood the importance of the rapid development of new technologies; men who promoted military-industrial innovations in such areas as aviation rocketry and tanks.

In many ways, military science and high technology were synonymous. Though the bright generals of Marshal Tukhachevsky's generation did not survive 1937, in many ways Stalin took the same approach in promoting applied science for the defense program. As students of physics, we had already heard that Beria himself, until his final arrest only a few months previously, had taken care of national security in both aspects: as the chief of security police, the KGB, and as the supervisor of high technology, especially in the nuclear area, for the country's military security.

I think I heard about the complex for the first time when we students were telling each other stories about different mailboxes. Despite the tendency of my friends to lean heavily toward going to such installations, my main argument against such places was quite straightforward.

Work related to the military represented not only the loss of physical freedom, such as the ability to travel abroad or even within the country, but also the loss of intellectual freedom. Instead of having the right to choose the kind of scientific problems to work on, you would be obligated to pursue the topics you were directed to work on by someone else.

Once, in the midst of our debates on future assignments, we were told that at a certain not very well known institution there would be an interview for those who were interested in nuclear physics. The name of the organization told us nothing. It was called the Laboratory of Measuring Instruments. I would say, as a potential theoretical physicist, I considered this name a real deterrent. But students often act as a tribe, and I decided to join my friends.

"Don't be silly," said one of my wiser and more experienced friends. "Don't take such a name too seriously. Even if it were called the 'Incubator for Chickens,' you shouldn't be afraid to go there and see what it really means."

Upon arrival on the distant outskirts of Moscow, we discovered that the facility's huge perimeter was enclosed by a stone wall. We could see only a few isolated buildings and trees inside the installation. We were not admitted beyond the reception room that served as the point of contact with the outside world. We were met by a bald man with an extremely energetic manner. He was clearly a sportsman, though his age was somewhere around forty, which, from my perspective, seemed unusual. I discovered his name was Andrei Budker. I had never heard his name before—certainly not among the prominent physicists. He briefly said that he was going to launch a very important and ambitious program in some, not yet identifiable for us, area of contemporary physics. He said he was looking for bright young students.

I was not particularly impressed. Despite his obvious charm and sense of humor, my preference was, once and forever, for Landau's school. So while my friends were seriously trying to explore this opportunity, I was idly following their conversation. I had no idea at the time that some years later destiny would bring me very close to Budker and his ambitious plans.

The interview was very different from the style of the theoretical Landau minimum. Budker asked questions that were rather simple and straightforward, although they required a quick reaction. I still remember one particular problem, which he asked with a kind of male bravado.

"Imagine for a moment," Budker said, "you are going to meet one of your two mistresses. You approach the metro station planning to take the first train arriving at the station no matter which of the two directions it is traveling. The lucky mistress would be the one to which you are delivered by this particular train. After many trials, statistics indicate that somehow one of the two young ladies would be visited three times more often than the other. How is it possible? Can you give an explanation quickly?"

I still was not interested in conversation, nor was I interested in the funny statement of this particular problem. Besides, I had never had the experience of finding myself in the situation under question. But somehow I rather immediately replied, "Oh, that's very simple. It has to be printed in a fixed itinerary for the trains that the intervals between the trains coming from different directions are uneven."

Despite my success with Budker's question, I was not much interested in joining this mysterious Laboratory of Measuring Instruments. I had made substantial progress with the Landau examination. In early 1955, still almost a year before the end of my university course, I had passed all seven of the theoretical "minimums" (with two more to go). Already considering the possibility of a quick breakthrough, I was thinking of asking Landau for the chance to do my diploma work under his supervision. Then an unexpected development suddenly changed all of my plans.

With a few colleagues of mine, I was invited by university officials to attend a meeting. In an atmosphere of solemn significance and secretiveness, we were told about our compulsory assignment to conduct our diploma work in an institution whose name was given to us only as a symbolic abbreviation of its address: Moscow Center 300.

But it was rather clear that something quite serious was behind this assignment. Even student gossip and speculation could barely help unveil the real nature of this place. I was quite upset at the thought that it could complicate my future job. I went to see Landau and told him what had happened. He became serious and said, "I am afraid I know where you have to go. This is a mailbox—one of the principal ones among them—where you will meet Andrei Dmitriyevich Sakharov. Have you ever heard this name before? He is an outstanding man. While I would not consider him a genuine theoretical physicist, his is rather a 'constructive genius.'"

A few days later a small group of graduates in theoretical physics boarded a special train that left late in the evening from the Kazan railway station in Moscow. With very little sleep, we woke up suddenly: the train was already standing in the corridor, between multiple parallel lines of barbed wire. It was the entrance to the place that Sakharov called in his memoirs "the installation."

The official cover name at that time, which was in use in conversations, was the "Near Volga Office." It was protected by a heavy-duty fence of barbed wire and included a huge area with forests, rivers, lakes, and a small town. The name of the town prior to its modern sinister predestination had been Sarov, which in its past had attracted crowds of pilgrims to visit St. Serafim's monastery. Its bell tower was still the most important reference point in the very center of the city. However, intensive construction work was seen everywhere.

We settled into a kind of dormitory set up in a contemporary apartment house. The next morning we went to introduce ourselves, according to the instructions we had been given. In a rather small, gray, three-story building, we were met with excitement and interest. Later we understood that in the brief history of the installation it was the first time students had been brought to participate in its historic mission. The Soviet counterpart to Los Alamos National Laboratory, the installation is now more commonly known as Arzamas-16.

The highlight of the first day was our meeting with Sakharov. I vividly remember his image. The impression he produced will stay with me for the rest of my life. A tall, slim man, he sat on a sofa in his office, his legs tucked under his knees. There was an air of serenity in his look. He was the nation's youngest academician, recently elected for achievements that were described only as "producing great service to the Soviet Union."

With a soft-spoken voice that was slightly burred, he told us what kind of topics for diploma work could be given. As I recall, he had a piece of paper with him on which he was writing what looked to be an outrageously complicated theoretical formula. But somehow my impression was that instead of formulas, in the corner of the sheet of paper there were also sketches—doodles—of airplanes dropping bombs.

My friends and I did not know, and would not even be allowed to know, that during this same period Sakharov and his colleagues were in a rush to prepare what Sakharov later called "the third idea": to design and produce a new "product." This product was to be completed and dropped from a bomber plane, in a critical hydrogen bomb test, by the end of that same year, 1955.

This test would bring to Sakharov further official recognition of his outstanding service to the nation, a second Gold Medal, and the title Hero of Socialist Labor. But it would also be the same test explosion that would irreversibly change Sakharov's thinking and set him on a new path, the path of international hero in search of peace and justice against thermonuclear terror and the totalitarian regime in his own country.

Sakharov told us during our visit to the installation that we were going to play an important role as guinea pigs. We were the first breed of students to be brought into the very heart of the Soviet nuclear program, where many of us would later pick up the torch, the flame of nuclear knowledge.

When the newcomers were assigned different supervisors, I was somehow given a lucky ticket. My diploma was not going to have anything to do with the nuclear program or nuclear weaponry. My problem involved a theoretical analysis relevant for astrophysical conditions of very hot matter in the state of plasma in stellar interiors.

The topic of my work originated in the discovery that had been made about twenty-five years earlier, related to what provides the almost eternal source of energy that makes the stars bright. Astronomers in the late nineteenth and early twentieth centuries had exhausted every possible idea that could explain the obviously inevitable energy crisis in stellar objects. They could not imagine that these hypothetical energy sources were any different from the familiar fossil fuels in terrestrial conditions.

All ideas were in vain. Even if God had pumped the best of oils to fill up the interiors of the stars, this energy would not have lasted long. Then Pierre and Marie Curie discovered the phenomenon of radioactive decay. Their work eventually led to early ideas about the much greater reservoir of energy contained in certain types of nuclei. However, it was not until 1920 that the idea first took intelligible shape. An eminent British astrophysicist, Arthur Eddington, suggested fusion reaction as a source of stellar energy: the bringing together of a few hydrogen nuclei to form a heavier one. This action would create an enormous amount of energy released as kinetic energy, the product of reaction. A very strong gravity force in stellar interiors would keep matter in a state of high density, thus providing the confinement of hot plasma required to keep thermonuclear reactions going.

Since Eddington's early pioneering hypothesis, many developments have taken place. Theoretical physicists have not only confirmed the principal possibility of thermonuclear reaction as a source of energy in stellar conditions, but perfected their theories to such an extent that, for any given star with known chemical composition and mass, it would be possible to calculate the energy balance. For our sun, all of these parameters were thought to be precisely known. Physicists constructed models of the solar interior with remarkable accuracy.

As my direct supervisor, theoretical physicist David Frank-Kamenetsky, one of Sakharov's team members, had been given this assignment. At that time he was undergoing a process of intellectual conversion from the "weapons kitchen" to open international science. I think the reason he became interested in astrophysical problems was the apparent similarity between bomb physics and stellar astrophysics.

In the office I was given a desk beside Frank-Kamenetsky and his assistant, who was doing mostly numerical computations. Both of them had already been personally converted into newly fashioned astrophysicists. At the same time, there was another young man, only a couple of years older than me, Yuri Trutniev. His duty was to represent Frank-Kamenetsky's small group in a think tank for weapons design. The system of secrecy was so overwhelming that no one even mentioned the atomic bomb. In that room, work was carried out at two opposite poles of this science: physics of extremely high temperatures and thermonuclear reactions.

Whenever Frank-Kamenetsky, Trutniev, and their occasional visitors needed to talk about their internal, presumably military, design issues they would use a jargon unintelligible to outsiders. The only thing that was clear was the word *device*. Innocent as it may have sounded, "the device" meant the ultimate weapon, the nuclear warhead.

As a threatening reminder of the place where we students from Moscow State University were doing our diploma work, one of my classmates, Valery Zolotukhin, had to deal with the impact of thermonuclear explosions on the environment. Poor Zolotukhin had to calculate how many hundreds or even thousands of tons of soil would be vaporized, not blown off, by the tremendous fireball of the H-bomb. Since this particular problem was not directly related to any specific design of the device, it was given the lowest level of clearance. From time to time our friend would share with us his doomsday findings.

The security and clearance regulations introduced in the life of that small provincial city, Sarov, were extremely tight. We students, before being sent to the installation, had to sign a special pledge to keep secret everything we knew in conjunction with our life in that place. All correspondence and letters were censored. For those who violated the strict rules, the government had a stick.

I remember in my time at the installation a trial of a young girl who, only a few weeks before, had been appointed medical doctor. She wrote a letter to her parents in which she tried to impress them with how significant and important her presence there was. She spoke of the installation as the capital of bomb design. Poor girl. Even if she had been intentionally willing to disclose secrets, she knew none of them. Still, the verdict was three years' imprisonment.

The deprivation of freedom for all the employees and inhabitants of this archipelago was compensated by privileges, bonuses, and special payments. There was a standard 20 percent salary increase for every employee. Thus, in the material sense our lives there were quite comfortable. We were well paid, and we lived in newly built apartment houses converted into an improvised dormitory. On the weekends we walked in the woods and swam in the lake that had been conveniently preserved inside the long barbed wire perimeter encircling our nuclear island.

In the autumn of 1955 our diploma work was over. We had to prepare for the last crossing of the border of this closed zone. I had a feeling of indescribable liberation from some unbearable gravity. I didn't know at that time that I was crossing the borders for the last time in my life and that my world line—my personal trajectory in time and space—was diverting forever from the world lines of most of my classmates and new friends in the Arzamas installation.

After five and a half years in university, we classmates were waiting for the last farewell party before each of us would make his or her own way in life. Every day a few of us would be invited by a special state commission to take an assignment and sign a contract. That night our dormitory would salute successful classmates with the corks from bottles of champagne. Then these classmates would leave forever the next morning. They were sad moments, but we tried to tell each other funny stories about our diploma practices at different places. We ridiculed the secrecy, elevated to cosmic scales, at different installations. We thought of them as the "animal farms" of the time, under the military-industrial complex.

It wasn't long before my class and I heard our sentence from the state commission. The message was unexpected and severe: a decree from the Council of Ministers, signed at the highest level, assigned the whole group of nuclear theoretical physicists to take compulsory assignments at a mailbox near Chelyabinsk in the Urals. It was, as I immediately understood, the same "installation number two," near Chelyabinsk, where some of my Arzamas friends, like Evgeny Avrorin, were moving. That explained the reason for the arrangements that sent some of us to Sakharov's installation for diploma work. The authorities wanted to see whether such guinea pigs were good enough material to join the staff of another mailbox to which the government clearly attributed unusually high importance.

From the very beginning we knew that we were in a no-choice situation. The government was very serious. We were faced with the futility of any kind of resistance. Most of my classmates signed the contracts.

I, on the other hand, was strongly prepared to fight. Landau, whom I had communicated with almost immediately on hearing the news, supported my fervor and encouraged me to stay in a state of internal rebellion. He promised to contact different important officials to ask for their help.

I waited in the students' dormitory among very few classmates in a similar state of rebellion. Almost a month had passed since I had launched my personal campaign of civil disobedience. From time to time I saw Landau or talked to him on the telephone. We discussed current developments.

After making some inquiries, Landau understood that the only person who could probably deal with such an issue decreed by the Council of Ministers was Kurchatov. Although I had never seen Kurchatov before, I had of course heard a lot about him. He was a legendary figure among physicists, known to us not only as an outstanding experimentalist in nuclear physics, but also as the driving force behind the Soviet nuclear program.

Landau personally knew Kurchatov quite well. By the time Landau approached Kurchatov, only two out of more than twenty of my classmates had been able to get off the hook. One of them was my colleague from diploma work at Sakharov's installation, who was released on the basis of his poor state of health. But it probably had not been done without help from his father, a prominent regional party boss. Another graduate was also released from this obligation, clearly with the help of his father's connections.

The rest of my contemporaries went to the new installation near Chelyabinsk with outright satisfaction—although they had no idea what kind of nuclear work they were going to do for the rest of their lives.

I knew quite well that material conditions at a place like Chelyabinsk would be much better. But that did not matter. I wanted desperately to avoid going. There was the moral side: I didn't want to work on a weapons program. Although Landau and I did not have any discussions about it, I think he understood my reservations completely. Landau took up my request before the historic changes began, spurred by the Twentieth Party Congress. That's why it was remarkable that Landau intervened at all. He essentially checked every level of authority, and only after he was unable to resolve my release did he go directly to Kurchatov to talk.

I was waiting for the final verdict, which was sure to come as a result of Landau's approaching Kurchatov. In many ways, I was certain of victory. Landau had very high prestige, and he enjoyed Kurchatov's respect.

Finally, a few days later, Landau told me that he had spoken to Kurchatov. Unfortunately, even Kurchatov was unable to release me from an assignment that had been given at the level of a governmental decree. But Kurchatov had been approached by other people with similar requests to liberate me from going to Chelyabinsk-70. The most Kurchatov was able to do was to ask for my change of assignment from the Chelyabinsk installation to his own institution in Moscow.

Landau's estimate was that this outcome was equivalent to an 80 percent success. "You will be here, available for my seminars," said Landau. "You can come at any time to discuss scientific issues with me." With mixed feelings I went to the university to sign the revised papers. This outcome played a big role in determining my future life.

7 / The Kurchatov Institute

➢

In mid-February 1956 I was given a university diploma and a letter of assignment instructing me to take a job at that enterprise with the routine and disappointing name: the Laboratory of Measuring Instruments, the stronghold of Igor Kurchatov.

On a snowy day, the public bus took me to the northwest outskirts of Moscow. Huge quarters were built up around a construction site, which had seemingly been developed in recent years. On arrival we were instructed to run between different bureaucratic offices to get through application formalities. The other draftees and I represented the huge spectrum of technical universities and colleges of Moscow. Quickly we formed a kind of horde invading and exploring a new continent.

Kurchatov's real estate was enormous. The centerpiece of this huge installation was the laboratory itself. It was surrounded by fortified concrete walls which were tightly guarded. Inside, the experienced eye could recognize the signatures of the nuclear reactors.

The first day I came to the institute I was given a bed in the dormitory. Typically, three or four people shared a room. It was definitely a step back after the privileged conditions that students of Moscow State University had enjoyed on the Lenin Hills.

A few days after my arrival I got an unexpected chance to move to another place that also was used as a dormitory. There was a line of a dozen cottages which were built during the first years of Kurchatov's installation. Each of them had a small piece of land, maybe an acre. Three people were put in each bedroom and about twelve to fifteen people lived in each house.

I had a place in this dormitory for more than three years. The cottages were considered much better than any of the other lodgings. I was told that the cottages were originally built after the war to provide better housing for German nuclear scientists and engineers who had been detained and forced to come to Russia and participate in the inception of the nuclear program.

I found a sad irony in this parallel: I, too, was a victim of conscription. I was brought to this place against my personal will. Even Landau's last words—that my compulsory assignment to Kurchatov's place had resolved the issue by 80 percent—did not give me much consolation. While my newly acquired colleagues and neighbors in the dormitory were openly rejoicing at their huge luck to be a part of Kurchatov's empire, I was unable to overcome the deep internal resentment I felt against the authorities who had deprived me of the free choice of a future profession. In the gloomiest moments, I compared myself to a kind of intellectual serf.

The Laboratory of Measuring Instruments had made an outstanding contribution to the creation of the Soviet nuclear bomb and nuclear establishment. Historically, it was the very first installation—a kind of early Soviet counterpart to Los Alamos National Laboratory—created by Stalin during the last year of World War II. The Soviets were aware of the prospects for the ultimate war instrument, the nuclear bomb, from almost the beginning of the Manhattan project.

In the cold winter of 1943, the government brought a few scientists and leaders of the scientific establishment and military industry to the Kremlin for the first meeting. This meeting launched the efforts to build a national bomb program. Kurchatov was chosen as the scientific leader of the bomb project.

Part of a new breed of Soviet scientists, Kurchatov was a pupil of the postrevolutionary period. He had had the chance to accumulate the wisdom and heritage of the old scientific culture during his years as a student of the famous academician Abram Ioffe. At the chief cradle of Soviet fundamental and applied physics—the Physico-Technical Institute in Leningrad—Kurchatov proved himself prior to the war to be a bright experimentalist in the newly developed field of semiconductors. Later he consciously decided to move to the area of nuclear physics, which he himself sensed would soon become even more important.

Among his scientific siblings, Kurchatov acquired the reputation as a man of great inner self-discipline and organization. A man of responsibility, he started even before the war was over to organize installation number two, which later acquired the code name Laboratory of Measuring Instruments.

From everywhere, the best physicists and the best engineers were recruited and assigned to Kurchatov's installation. The nuclear bomb program was given the right of extraterritoriality. Anything requested by the "saviors of the nation" would have to be delivered as soon as possible, despite the difficulties of the late war and early postwar years.

As huge as it might seem to the occasional visitor, Kurchatov's installation, to which I was assigned, was only a cap, the smallest upper tip of the huge iceberg that consisted of dozens and hundreds of different nuclear enterprises. But it was the most important part of that quickly growing nuclear empire, which later was given the trivializing name the Ministry of Medium Machine Building. With the Laboratory of Measuring Instruments at the top of the enormous scientific and industrial iceberg, there were a few more centers of intellectual activity, created in parallel, which performed specific tasks. Sakharov's place, the Arzamas installation, for instance, which was already familiar to me, was responsible for the ultimate bomb designs.[1]

Such a massive concentrated effort brought the desired fruit. In 1947 at his own installation in Moscow, Kurchatov achieved the first important breakthrough. His team put together the first reactor, the assembly of critical mass, an achievement that paralleled the design of the first reactor at the University of Chicago in 1942 by Enrico Fermi's team. The designers of the first nuclear bomb were supplied with all necessary data and materials. In 1949 Stalin got a report on the successful test. The final assembly took place in the Arzamas installation.[2] The fission bomb tested in 1949 was designed by a team led by Yuli Khariton, the scientific head of the installation from the very beginning.

After the distant fireball vindicated the efforts of Kurchatov's team, Artsimovich, one of the most prominent members of the Kurchatov-Khariton team, was heard to make the dry remark: "Imagine how bloody this test would have been if the bomb hadn't gone off."

In mid-February 1956, I arrived at my destination inside the huge Kurchatov campus: a one-story building with the name Bureau of Elec-

1. Only recently was the name of Sakharov's installation officially disclosed under the title All-Union Research Institute of Experimental Physics.

2. Sakharov, however, was not part of that project. He was a newcomer for the follow-on project, to deliver an even more destructive ultimate weapon—the hydrogen bomb.

tronic Equipment. When I heard the name, I felt absolutely helpless. Even my decision to become a theoretical physicist was to a certain extent motivated by the complete absence of any talent with technical equipment, of which electronics would be the most sophisticated and inaccessible for an abstract-minded theorist such as myself.

Upon arrival I was introduced to another recent recruit, Taras Volkov, who was given the task of introducing me to the Bureau of Electronic Equipment. While we were exchanging the first sentences, I was taken by surprise by something that sounded like cannon fire or the explosion of grenades. It came in intervals, with the same sharp, strong sound. Volkov smiled and said this was "it." "This is what we are doing here."

"What?" I asked.

"Fusion," he said.

This was precisely the new part of the Kurchatov Institute that was established after Lavrentiev, my roommate in Stromynka dorm, sent that famous letter to Comrade Stalin suggesting the brilliant idea of how to control fusion. So my new job was created in some way by Lavrentiev! I asked my interlocutor if Lavrentiev's name was familiar to him. He said, "Of course, everyone knows him, but he is not here anymore."

The principal focus of my new institution was briefly described as the study of plasma, the medium in which atoms of heavy hydrogen are heated to substantially high temperatures. The use of electric discharges, created as a by-product in these explosions, was thought to be the simplest way to produce hot plasma.

"Now I have to introduce to you some very specific and strict rules, which will protect the secrets of how we will make fusion," said my new friend. "We use a special language to communicate."

I was quite overwhelmed at the thought that I had to learn another language. It was a special one, a kind of Aesopian tongue. But this time it was created not by conspiratorial revolutionaries, but by Beria and his people.

The whole issue of controlled fusion—and, as I could expect, uncontrolled fusion required for the bomb—was considered top secret. All the scientific wisdom accumulated during the scientific revolution of the twentieth century and finally published in hundreds of scientific papers was brought together to be translated into a nonexistent language, invented only to mislead potential spies. I still vividly remember the ingenious inventions of Beria's philologists. *Neutrons,* the final product of fusion reactions (the very appearance of which would signify that, yes, high temperatures had been created and the reaction is going on) were called "zero points." *Plasma*, a very decent and inoffensive name, was changed to "syrup." Of course, the utmost secret was to talk about heating "syrup" to extremely high temperatures. Clearly, no spy could guess the meaning of a call to bring "syrup to the highest altitude." Thus, the subject of my studies was to be "high-altitude syrup."

Apparently, at the beginning of the decade, the controlled fusion team at the Kurchatov Institute thought that they were very close to a breakthrough in energy production. Unfortunately, it turned out to be nothing but self-deception.

The very idea of using powerful electric discharges had seemed very smart. The experiments were performed in special vacuum vessels filled with heavy hydrogen. Two electrodes were used to apply high voltage to the gas. The application of the voltage would cause an electric breakdown, which formed a strong electric current flow through the gas, causing the gas to undergo rapid ionization. This is how the plasma (sorry, the "syrup") was produced. The magnetic field needed to confine it, according to Sakharov's and Tamm's invention, would be automatically in place as an indispensable companion to the electric current.

Early experiments with the powerful electric discharges were tried from 1951 until the time I joined the Kurchatov Institute. Magnetic fields generated in them were strong enough not only to provide the desired confinement, but even to compress the plasma (to pinch it), thus forming a narrow cylinder of plasma.

Success came rather soon. In late 1953 and early 1954, the cannonade of electric impulses created the first neutrons. Physicists in the lab were proudly whispering to each other, "Neutrons," and then speaking loudly, "Zero points."

However, the state of euphoria soon evaporated. Despite expectations, the further increase in voltage did not lead to a rise in temperature or an increase in neutron production.

By the time I came to the lab, they already knew the prognosis: unlike a stable sweet syrup, the plasma as a physical substance is extremely restive and peculiar. *Instability* became the key word to describe this quality. Because of that, the process of pinching would never preserve the regular cylindrical form. The plasma would find ways to escape, carrying lots of invaluable energy with it to the walls of the discharge device—and dissipating chances of a thermonuclear reaction.

But what about neutrons, which were already detected in early experiments in the process of pinching? Were they not the messengers of the thermonuclear fusion reaction? Alas, my colleagues came to a painful and sobering conclusion: the neutrons resulted from processes that had nothing to do with high temperature. They resulted from a small number of deuterium nuclei accelerated by an electric field via a mechanism completely alien to genuine heating. Such a mechanism was an artifact. It was capable of evolving into fusion reactions, with only a very small fraction of nuclei. There was no chance of developing it into a thermonuclear reactor.

In that sense, the Aesopian language as a weird invention gave an accurate prediction: these neutrons were nothing but zero points.

8 / My Mentors

➢

My first reconnaissance visit to the Bureau of Electronic Equipment brought me some consolation. I would be working on peaceful controlled fusion instead of having anything to do with the nuclear weapons program. But even such a benign formula did not liberate me from feeling that I was playing the role of a pawn, an intellectual serf.

My new colleague Volkov, who had introduced me to our secret, was friendly and sympathetic. However, he and the others were still very different from the bright wunderkinder I had hoped to be surrounded by in the inner sanctum of the Landau theory school.

During the next few days after my arrival at the Kurchatov Fusion Plasma Lab, events seemed to develop at cinemagraphic speed. I was learning more and more, not only about fusion, but also about my future colleagues.

Mikhail Leontovich, my boss, was the head of the theory division of the bureau. His was not a new name for me. He was one of a handful of Soviet theoretical physicists who had been raised to the level of full academician. His name was mentioned during the student rebellion at Moscow State University as someone who should be invited back to teach contemporary physics.

I will never forget my first real introduction to Leontovich. He entered the room with a cigarette in his hand, preceded by the sound of very strong coughing. From that very moment, I almost never saw him without a cigarette; his trademark cough was always the precursor to any entrance.

Whenever he came to confer on a subject, his first question was always: "What is going on here?" Then he would listen for hours to long briefings on a problem. I have never known a supervisor, or a teacher, who spent so much time with each pupil. Leaving aside the commentaries and interpretations, he would sit next to you and actually help resolve the mathematical difficulties you were encountering.

Leontovich was a natural caretaker, ready to sacrifice his time and ambition to help his pupils. When I met him for the first time, he had already decided, at the age of fifty-three, to abandon his own personal plans to do research by himself, to write scientific articles, or even to give invited talks on scientific topics. All of his time was consumed by one overwhelming task: to help his team in physics. That was the basis of our relationship with Leontovich.

I found it more and more interesting and exciting to communicate with him every few days on my findings. Clearly, his way of thinking, his mentality in theoretical physics, was very different from that of Landau and his school. I wouldn't say he was very quick by Landau standards, but he was very fundamental, very solid. And he had a great talent for interpreting complicated formulas. He had an intuition, or as we would say in the company of physicists, an ability to express everything in "the language of a few fingers."

In the tradition of the old Russian intelligentsia, Leontovich was very democratic and accessible. I spent many hours in his apartment, which exuded the semispartan atmosphere of simplicity and friendliness. Sitting in the kitchen, we had discussions on almost every topic while his old maiden aunt, Marpha, would serve us only one dish, buckwheat porridge.

Leontovich was interested in everything that was going on in my life and the lives of my colleagues. "Do you have all the material logistic problems in your dormitory resolved? Do you need any kind of protection, promotion?" He was ready to defend our interests against the bureaucratic apparatus of the lab or the institute.

In the closed circle of friends and pupils, Leontovich's behavior was awarded with the affectionate title "Academician Hooligan." Leontovich proved many times that he really deserved this unofficial title. In the summer of 1964 he was one of the first to publicly attack Lysenko and his protégé, Nikolai Noujdin, at the huge official meeting of the Academy of Sciences that had convened to elect new members.

With eloquent speeches, Leontovich and Tamm managed to change the minds of the audience by denouncing Noujdin as a "forger of true science." He was rejected by the majority of academicians.

The incident was almost immediately reported to Khrushchev. Apparently, the news absolutely outraged him. "We don't need such an academy," he is said to have remarked.

If the principle of natural sciences that says that "action will create counteraction" is applicable to human relations, Leontovich certainly was a showcase for that. The admiration shown by his friends and colleagues was accompanied on the other side by an official skepticism and caution.

Years later I heard a story from Sakharov about the circumstances under which Leontovich himself was appointed to the Kurchatov Institute. Very soon after Lavrentiev's letter to Stalin and Sakharov and Tamm's subsequent support and modifications of his idea, the government launched an ambitious project to materialize these ideas about controlled fusion. Kurchatov opened a new division in his institution.

At a special meeting chaired by Beria, the issue of who would be given the very important role of head of theory for this new project was raised. Tamm and Sakharov were busy with the H-bomb project, but both of them knew Leontovich very well and they suggested his name. There was clearly overwhelming support for Leontovich on the side of the scientists and experts. According to Sakharov, during the meeting, unexpectedly—at the very last moment—Beria's assistant silently passed a clip of papers to Beria. Beria quickly looked at them, then remarked: "Maybe it is so, but it only means that we must watch him more closely."

Being once accused by authorities of listening to the broadcasts of "enemy voices"—Leontovich made no secret of following Voice of America or the BBC—he made a sarcastic remark: "If a dog is not fed by his patron, it will pick at someone else's slop pails!"

Leontovich maintained his boyish rebelliousness and independent views until his death. Despite this, however, he never actually crossed the invisible frontier that divided courageous (but "still normal") behavior from the "dissidence" of those who actually burned their bridges. Nonetheless, Leontovich was close to many who did. For instance, he helped the family of Boris Pasternak, who was the subject of real ostracism and a massive boycott not only by the officials in the government, but also by the vast majority of his literary colleagues.

Very few people had the courage to openly support Pasternak when Khrushchev himself expressed his utmost displeasure that the Nobel Prize had been given to Pasternak's *Doctor Zhivago*. Though Zhivago was not a dissident, his fate exemplified the fate of the Russian intelligentsia who tried to keep their integrity, decency, and honesty through the terrible, turbulent years of Russia's tragic history.

In a very strange way, for me Leontovich had always personified the image of Doctor Zhivago, with his nonviolent but unconditional and firm rejection of the totalitarian regime.

The real strength of the Kurchatov team in fusion came from very intimate cooperation between theory and experiment, where academician Artsimovich established himself as the undisputed leader. Artsimovich had been educated at the same Leningrad school of Abram Ioffe, from which Kurchatov and many other prominent Soviet physicists came. At the end of the war, when Artsimovich was rather young, there was a need to recruit bright people into the Soviet nuclear establishment. Kurchatov, of course, knew him very well. Artsimovich was assigned to design a special technique to separate uranium isotopes to build a first stock of uranium 235, the deadly one.

Artsimovich rarely spoke of his own contribution to the nuclear bomb project.[1] One of his duties at the beginning of it, as I understood, was to supervise the Germans. Probably it was from them that he learned the basics of early prewar plasma physics. It was almost completely irrelevant to the bomb program, but it became very important in controlled fusion. Artsimovich clearly respected at least some of the German "guest" workers in the Soviet nuclear establishment, but he attributed to Klaus Fuchs a much bigger role in helping the Soviet bomb program. On the same issue, Sakharov remarked many years later that in his estimation Fuchs's intelligence was of secondary importance. The principle nuclear secret, he said, was the knowledge that the nuclear bomb could be devised and exploded. In that sense the secret was known in advance of the final Soviet breakthrough.

The theorist and the experimentalist, especially in physics, are inseparable for the progress of scientific knowledge. But the approach, the logic, and the technique—even the mentality—are completely different. These differences created the framework for a lot of chafing jokes and a lot of serious debates. Which of these two types was more important for unveiling the secrets of nature? Of course, we young theorists were proud of such titans as Einstein and Bohr. And after all, the international reputation of Soviet physics was based on respect for Landau and a few other prominent theorists.

1. Though it was known that Artsimovich worked on isotope separation.

Artsimovich's dual talents helped us a great deal. He was one of very few experimentalists with whom theorists could speak in their own language. In a funny way, I believe that this type of ability is closely coupled with a deep sense of humor and intuition. I consider his definition of what our fraternal colleagues in high-energy physics were doing as the best: he used to say that high-energy physicists, who build accelerators that are ever-growing in size and budget to try to understand the internal structure of elementary particles, reminded him of extraterrestrials trying to discover the secret of auto making by colliding cars at very high speeds and then studying the debris. Alas, nobody has yet suggested a nondestructive technique for such studies of the elementary building blocks of matter.

Artsimovich's contribution to the big science of plasma fusion is substantial and undisputed. Moreover, his role was a very important moral factor in the whole international controlled fusion program. For quite a few years, almost from the beginning of that research, there was a widely recognized syndrome called Bohm diffusion. On the basis of early experiments with plasma confinement in magnetic fields, many researchers came to the rather pessimistic conclusion that there might exist a universal physical mechanism, an incurable disease, that would always destroy the confinement of hot plasma by the magnetic field. Artsimovich launched systematic experiments with a device originally suggested by Sakharov and Tamm, a plasma torus now universally known as tokamak. These studies were capable of completely eliminating Bohm's syndrome and the defeatist mood among the fusion community.

In the summer of 1956, a few months after I joined the Artsimovich lab, there was a huge celebration. Both of my mentors, Artsimovich and Leontovich, along with four other scientists, were decorated with the prestigious award the Lenin Prize. At the banquet, Artsimovich delivered one of his most brilliant toasts. Referring to the larger group of scientists and engineers who were an important part of the whole project, he said, "Science nowadays is a collective undertaking."

Artsimovich also had a passion that fed his soul. He was deeply interested in geopolitical issues. Eventually, he fulfilled his secret dream: the desire to pursue strategic thinking. In the mid-fifties, he joined the international movement of scientists, the famous Pugwash conferences. The best scientific brains applied their minds to the thermonuclear deadlock. Throughout the most difficult periods of confrontation—the ups and downs of the Cold War—the Pugwash meetings remained the only reliable channel for important arms control discussions between the Soviet and American blocs.

Artsimovich soon became one of the most prominent figures on the Soviet side. Among his interlocutors on the American side were George Kistiakowsky and Jerome Wiesner, presidential science advisors in different administrations of that epoch. Even Henry Kissinger, when he was still a professor of political science at Harvard, was a participant in Pugwash brainstormings. It is from that epoch that we inherited the apocalyptic notion of MAD (mutually assured destruction) and the concept of nuclear overkill. Artsimovich and other thinkers of his generation were the founders of the science of strategic stability in the nuclear age.

No matter how great were our mentors, we learned a great deal from our contemporaries, too. At the Bureau of Electronic Equipment, I spent hundreds of hours working with a collection of outstanding young men. A close friend of mine, especially during the first years at the Kurchatov Institute, Vitaly Shafranov, was one of the young scientists who shared with Artsimovich and Leontovich the famous Lenin Prize for early works in controlled fusion. His theoretical analysis produced the first robust recipes of how to keep plasma more or less stable. Even though it was later understood that it did not serve as an absolute cure, it was at least the first step in making plasma controllable. It still bears the name Shafranov criteria.

Many years later, when the time came to succeed Leontovich as head of the theory division, Shafranov was certainly the best choice. Even if his scientific interests were not as diverse as those of Leontovich, his approach was always solid and competent. Under Shafranov's leadership, Kurchatov Plasma Theory is still one of the leading groups in the world today.

9 / Kurchatov

➢

At the time I joined the Kurchatov Institute, I didn't realize that Igor Kurchatov had a vision far beyond the issues of nuclear parity. After many years of running the nuclear military program and attending dozens of nuclear explosion tests, he had achieved the goal of delivering a nuclear defense, as he saw it, into the hands of the nation. It was then, I think, that his psychology began to change.

In the last years of his life, Kurchatov was completely absorbed and even obsessed with the idea of developing something very different—something peaceful perhaps—for which he could be remembered. I believe that he had a kind of guilt complex from his many years of work on the nuclear program. Whatever it was that motivated him, I can testify it was a real passion. He persuaded the government to support the construction of a nuclear reactor for civilian use, and very soon this branch of the Kurchatov empire had developed into a powerful independent institution. Nothing at that time could have provided any hint of its future: the catastrophic failure of the Chernobyl reactor.

Kurchatov strongly supported the development of high-energy accelerators to keep Russian science on the frontiers of contemporary physics. To achieve this goal, the government financed the construction of Dubna, a huge campus sixty miles north of Moscow on the Volga River. Even before this fairly expensive and fancy nuclear installation was put in the hands of an international scientific consortium comprising scientists from Socialist bloc countries, quite a few outstanding physicists with international reputations were brought to that place.

Surprisingly, Bruno Pontecorvo, a bright Italian nuclear physicist and a former pupil of Fermi, came. It was hailed as a great event, as a reverse brain drain. Prior to coming to the Soviet Union, Pontecorvo had been a member of the British Atomic Energy project. His defection to Russia followed the arrest of Fuchs, "the atomic traitor." That coincidence created a lot of debates over the potential association between Fuchs and Pontecorvo. However, upon arrival in the Soviet Union,

Pontecorvo had nothing to do with classified military projects and, under the auspices of Kurchatov, soon became one of the leading high-energy physicists.

The high-energy physics community eventually constructed its own empire. Several big machines were built in the country. At times there were even world record holders. But something was elusive in a society whose principal motivation for supporting science was to strengthen the power and prestige of its military. Very few prominent results were obtained; very few experimental successes were realized. While disseminating the seeds of nuclear science and technology at different locations in the country, including almost obligatory nuclear reactors in the capital of every ethnic union republic in the country, Kurchatov kept for himself, for his beloved Laboratory of Measuring Instruments, the project he considered his swan song: the controlled fusion program.

Soon it became clear that Kurchatov had a lust for this idea. The political horizon gave Kurchatov the chance to promote his grand vision. In the summer of 1956, he was invited by Khrushchev to accompany him on a very early foreign trip to the United Kingdom.

However, the centerpiece of Kurchatov's visit to England was not as junior companion to Khrushchev or Nikolai Bulganin, then the prime minister. He had a separate invitation to Harwell, the sacrosanct cradle of Britain's highly classified nuclear program, which, until Khrushchev and Bulganin's visit, had opened its doors only to Fuchs and Pontecorvo as foreigners.

However, Kurchatov hadn't come to Harwell to learn about Western nuclear secrets; he had come to disclose his own. In a sensational unprecedented talk, he described the controlled fusion concept being developed in the Soviet Union. A large part of his talk was dedicated to the description of the experiments with powerful electric discharges in plasma that gave rise to the production of neutrons, originally thought to be messengers of the thermonuclear fusion reaction and later interpreted as artifacts.

These circumstances gave skeptics a chance to interpret the whole move, by the official government and Kurchatov, as a bluff. There was no way, the skeptics said, for them to have developed an immediate fusion reactor, and since the Soviets had insisted that all these experiments with fast discharges were a fake, the Russian government had nothing to lose by disclosing such "secrets."

There might have been some truth to this argument. However, the logic of international science would not necessarily require all scientific disclosures to come only when the technology has been fully developed. If you present a premature idea, and if there is true promise in it, it should soon create a chain reaction of research activity involving many nations and institutions.

Kurchatov delivered one of the top Soviet secrets, the very birth of which was supervised and directed by Beria. However, it was not a secret for American and British scientists. A few years earlier, they had launched a program of controlled fusion of similar dimensions to Kurchatov's own. They had their own engineers, young inventors equivalent to Lavrentiev, and their own leaders, capable not only of getting neutrons in powerful electric plasma dischargers, but also of understanding that they are the artifacts of peculiar acceleration processes. They, too, knew that the road to achieving a stable controlled fusion was a long one.

At the Bureau of Electronic Equipment, the think tank of controlled fusion at Kurchatov's own installation, we felt the impact of Kurchatov's disclosure very quickly. The ridiculous classified language was immediately abandoned. Kurchatov came back from England and opened his own regular seminars, which soon became the most influential and prestigious gatherings in Moscow. There we discussed different ideas and scenarios of how to achieve a controlled fusion reactor. This is how I met, for the first time, many of those whose names were usually spoken only in whispers, such as the name of weapons designer Yakov Zeldovich.

At the first few meetings, we young theorists of Leontovich's team occupied the rear benches in the large room. Mesmerized by the names of prominent people, we followed the speeches. However, rather soon a revelation came: high-temperature plasma was a new science, and it was we, the young people, who were the experts in plasma, not the older, more experienced scientists. I did not quite realize it myself until I summoned the courage to jump to the podium to disprove, with boyish ardor, what I thought was a rather naive suggestion by academician Vladimir Veksler, the man who was considered a classicist of the then-biggest accelerators and the recipient of the American Atoms for Peace prize. Kurchatov remembered my bravery, and not long after the meeting I was invited for a lengthy private conversation with him.

As a result of Kurchatov's campaign for controlled fusion, the monopoly of the Bureau of Electronic Equipment was broken. Kurchatov launched several new experimental and theoretical groups in Moscow, Kharkov, and Sukhumi.

Artsimovich, one of the first beneficiaries of the grand opening of fusion science, made his first trip abroad a few months after Kurchatov's triumph at Harwell. He went to Sweden to encounter, for the first time, an international audience of plasma physics and fusion professionals. Everything was new to him and his Soviet colleagues. Sending the Soviet delegation was like forming a team to explore an undiscovered continent. The authorities did not know how to let Soviet nuclear scientists meet the world. At the seminar upon his return, Artsimovich joked that he had introduced his indispensable companion, a bodyguard, as "Professor Logunov." The hosts were too polite, he told us, to embarrass the poor fellow with questions on professional topics.

Following Kurchatov's initiative, the Western world slowly started to release declassified papers on plasmas and fusion. We read them, studying them as long-awaited signals. It was not so much that they delivered scientific revelations; most of the ideas were quite familiar and paralleled our own research. The most important message was that we were not alone in the fusion science universe. There were people of the same breed, of the same family of international science, out there. That feeling created a strong urge to meet our counterparts face to face.

In the summer of 1958 we were told that one of the most prestigious international conferences on the peaceful uses of nuclear power would meet in Geneva. Enthusiasm was overwhelming when we were told that for the first time controlled fusion would be an indispensable part of the program.

The news about participation in the international Geneva conference went through Kurchatov's empire like a shock wave. Everyone was excited, trying to guess what our chances were for inclusion on the Soviet team that would take part in the unprecedented event. In the meantime, the immediate issue was to declassify the long, thick volumes of reports in that funny Aesopian language. Once while doing this job, I got an unexpected phone call from Kurchatov's secretary. She said that Kurchatov urgently wanted to see me. I was completely unprepared for such an immediate appointment at the top and went to find Leontovich. He listened with a funny smile and said, "Okay, you have to go. We can probably cook up something interesting out of it."

The meeting with Kurchatov took place in his studio on the first floor of a rather substantial Russian cottage. Kurchatov was extremely friendly and casual. He even joked sometimes, touching his remarkably rich, long, gray-flecked beard. He looked like someone straight from a traditional Russian fairy tale. His dark beard seemed to symbolize the roots of Kurchatov's dedication to his country's wealth and security. But our conversation was not about a weapons program. Kurchatov wanted to ask me what I thought about the most difficult issue, that of the controlled fusion reactor.

For a few hours, we talked and I told him about different types of dangerous instabilities that could put at risk the construction of plasma devices, the final goal of achieving confinement of hot plasma. He took copious notes, putting everything I said in a huge notebook in his very accurate and condensed handwriting. From time to time, Kurchatov asked technical questions and even referred to the statements or opinions expressed by his other invitees. I understood that it was an enormous honor for me to be part of such a private conversation, since the names of the predecessors he quoted were extremely impressive. As a matter of fact, Landau was among them. I have never been involved in such a briefing in which I have been asked questions with such a genuine interest, all of which were right on target.

At the conclusion of the meeting, I went running back to see Leontovich to tell him what had happened. He thought for a couple of minutes and then declared, "You, my friend, are going to Geneva! Now, I will make sure it happens after this very important meeting."

The bosses had their own agenda for preparing us for the conference. We had several briefings and, finally, the whole delegation was called in for a concluding instructional meeting with one of the highest officials in the Soviet nuclear establishment. It was Dmitri Yefremov, an engineer who had spent most of his life in different branches of Soviet military industry, until he finally rose to the level of deputy to the minister of Medium Machine Building. Yefremov's final advice was: "The main focus for your activity at the conference should be to learn a dollar's worth of science and show a kopeck's worth in return."

I was startled by such a statement. How can we establish our leadership in international science and technology with such an approach? Wouldn't such a position make your activity worth only a kopeck after all?

A few days later, on a warm August day, "the world's first passenger jet," the Tupolev 104 from the "classless Soviet society," carried our group to the Geneva airport. My friends and I had mixed feelings. We knew we were going to one of the most beautiful countries in Europe—at least it looked like that in the postcards. At the same time, we knew that it was our first trip ever to a world divided by conflicting classes. That's why we were ready and prepared to be vigilant.

The reality we encountered was not simply charming, it disarmed us. As we encountered people walking along the streets of Geneva, we experienced a strong shock. They were without the imprint of fear on their faces. They looked prosperous; indeed, some looked wealthy. Everything, including the Lake of Geneva, the fountains, and the mountainous landscape, radiated peace and stability.

We started asking each other "Where is the oppressed working class? Where are the proletarians?" Someone finally suggested jokingly that maybe Lenin had experienced such a shock when he came to this pastoral, bucolic country looking for political asylum. Perhaps it had created in him an irrepressible urge to go back to Russia as soon as possible to start a revolution. Fifty years after Lenin walked along the streets of cozy Geneva, we were asking probably the same questions that had crossed his mind. When, at last, would Russia catch up? When would it join the civilized world? When would Russia have enough wealth to move people from communal apartments to decent housing?

As the youngest and most inexperienced in our group, I made a very naive and overly optimistic guess. Five years, I said. Shafranov, a few years older and wiser—after all, he was a Lenin Prize winner already—suggested ten years. However, we had a real veteran among us, Mikhail Ioffe, one of the brightest experimental physicists at the Kurchatov Institute. As an experimentalist, he was more practical by definition. Besides, he had spent several years in the trenches in World War II. He looked at us with an ironic smile and said, "No, boys. It wouldn't work so soon. It would take at least twenty years to bring a similar prosperity to the Soviet Union."

How naive, how simplistic we all were at that time, imagining the beautiful tomorrow, the economic wonders of socialism.

Still in shock after the first encounter with the capitalist world, we were also going to meet for the first time with our counterparts, the American controlled fusion team—mirror reflections of ourselves. The meeting place was a huge topical exhibition titled "Atoms for Peace." It was built as a colorful and exciting feast of creativity and invention that visualized, for the layperson, the concept of different controlled fusion devices. We all gathered around the exhibition. Improvised meetings next to the Soviet or American expositions quickly brought us together.

We first-time travelers expected Americans to be like extraterrestrials. To our surprise, we met a group of modest and similarly embarrassed young people. After a few moments of reservation and restraint, we finally came together to discuss scientific issues. We understood each other. It was like a miracle to meet extraterrestrials who understood your own language and to be able to follow what they told you.

Long before the start of the official conference, we had already established the backgrounds of our scientific work. The most remarkable revelation was that, without any contact and on different continents, we were in essence doing almost the same things. We, independently of each other, had rediscovered not only the principles of hot plasma physics relevant for controlled fusion with its instabilities, but even specific inventions to keep plasma under control, using sophisticated configurations of magnetic fields. In many instances, we were almost in parallel with each other.

The formal conference was almost unnecessary from the point of view of learning anything new. The particular session at which I was speaker was chaired by a famous Indian physicist, Homi Bhabha, and a Soviet physicist, Tamm, the teacher and mentor of Sakharov. Tamm was also already known as coauthor of the idea of magnetic confinement for fusion. I remember the discussion he initiated on the prospects of controlled fusion. The participants tried to be very cautious in their prophecies. Bhabha said that a controlled thermonuclear fusion reactor would be an unusually sophisticated and difficult undertaking, which might even take up to twenty years to develop.[1] Many of us thought

1. It would be very difficult to find a serious scientist or engineer now, thirty-six years after the Geneva conference, who would say, "Yes, it would take us twenty years to build a fusion reactor."

Bhabha's prediction was too pessimistic—in fact, almost as gloomy as Ioffe's prophecy on catching up with Western civilization.

Discussing critical plasma problems with our newly acquired friends was much easier than discussing political issues. We tried to avoid them as much as possible. Later some of our new friends told us, "You fellows are too timid! You're almost uncommunicative!"

At that time, we children of the Twentieth Party Congress, had developed a peculiar way of expressing our political views and assessments through satire. This political satire was encapsulated in dialogue on an imaginary "Armenian radio" and sounded like: "What is the difference between capitalism and socialism?"

"Well, if capitalism means that one man exploits the other, socialism is exactly the opposite."

This type of joke was the limit of our internal exile. And maybe not without reason. Shafranov, Boris Kadomtsev, and I discovered at the Geneva conference a person in the Soviet delegation who was never too far from any kind of conversation we had with our new American friends. The moment we introduced him, reading the name from his badge as "Professor Konstantinov," we understood that his English was much better than ours and that his linguistic expertise was much superior to his knowledge of physics, let alone controlled fusion. However, his very "employ" became clear one evening when he followed us out on the town.

We had an appointment to go to the cinema with the young American plasma theorist Martin Kruscal. At the entrance to the movie theater, I found myself so embarrassed by the newcomer's presence that I hastily bought tickets for all of my cohorts. While waiting for the movie to begin, our uninvited comrade started to provoke Kruscal with rather offensive conversation.

Kruscal asked him, "Where did you learn such good English?"

He answered, "I spent a couple of years in a special school."[2]

This statement was absolutely shocking for all of us in the Soviet trio. We also quickly realized that this guy must be absolutely drunk. Konstantinov relentlessly continued his political remarks, trying to humiliate our American colleague. Entering the movie hall, we tried to separate our American guest from this obnoxious drunkard. Then, after

2. In the Soviet Union, this meant only one thing—a KGB school.

the movie was over, we decided that further interaction was simply unbearable. We put Kruscal in a taxicab and confronted our compatriot. Konstantinov was absolutely mad with anger.

"Why did you let him go? We had to beat him up!"

Long after midnight, Kadomtsev, Shafranov, and I were still discussing every detail of the evening's party. We were depressed and terrified. Finally, we concluded that this guy was indeed following us and, at a certain moment after drinking too much, he had lost control.

The main issue we discussed was what to do. Should we keep silent and put at risk not only our new friendships, but also the prospect that this guy might commit more serious offenses? Finally, we decided to tell the whole story to Leontovich, a wise man who would tell us what to do.

The next morning, Leontovich evaluated all of the details of the story and suggested that the most proper behavior would be to explain everything to one of the experimentalists on the controlled fusion team who was also the leader of local party organization at our Bureau of Electronic Equipment. He was considered a rather decent person, having had many years of military service behind him in World War II. He was also a confidant of Artsimovich, Leontovich's choice in a situation where the presence of the politcommissar was unavoidable.

Mikhail Romanovsky carefully listened to our story, and then he said he would take the responsibility of talking to someone at the very top of the delegation.

The call from the top came within a very few hours. We were invited by Romanovsky to visit the person he had described as the most influential in the delegation. We were led into a monumental suite in the same Metropol Hotel, on the promenade overlooking the Lake of Geneva.

Our host was a man I had never seen or heard of before: Ivan Serbin. He did not introduce himself, and we were smart enough to understand that this man was the gray cardinal of the Soviet group here. Though he was the boss, we could tell he was not one of the academicians or even one of the ministers. He listened to our story dryly, without irritation, which made us feel much more courageous.

I had been deputized by my colleagues to be the principal speaker. Describing the story, I tried to stay calm, stressing mainly our great shock and surprise that anyone could deliberately undermine the very early stages of building international scientific cooperation, especially after such an important event as Khrushchev's trip to the United Kingdom.

Serbin thought for a while and then he spoke: "Don't worry. Continue your duties at the conference. Nobody will undermine your efforts in the international arena anymore." He rose from his chair, indicating that the conference was over.

We immediately retreated and tried to understand whether it was an indulgence or a blessing coming from the man whose precise position in the hierarchy Romanovsky had been explained to us was head of the Defense Department of the Central Committee of the Communist Party. Finally, we concluded that in a kind of perestroika launched by Khrushchev, the role of the KGB as the supervisor of strategic sciences and high-tech development in the country was gradually passing to party apparatchiks. In fact, Serbin was doing almost the same job that until a few years ago had been done by Beria.

We didn't have to wait long for the results of that meeting. By lunchtime Konstantinov, seemingly embarrassed, approached us and said a few words, trying to apologize for his misbehavior or the misunderstanding. No one could tell what he really had in mind. However, with these words he handed us a small sum of money, a few Swiss francs, as reimbursement for last night's tickets to the movie. That was all. But it was a symbol of our victory.

The rest of the conference was even more exciting. Liberated from continuous surveillance, we returned to our own circle where we belonged. We again debated the physics of fusion and made our first contribution to future long-lasting international friendship. We debated every potential disagreement. Objective truth would be the winner, we thought, and at the same time the victory would be shared by everyone. Leontovich, with his modest, unpretentious style of debating, quickly won moral authority and the respect of our new friends. However, extremely ambitious and combative, Artsimovich was not ready to agree easily with the counterarguments of his foreign counterparts. Sometimes it was painful for him to admit that he missed certain interesting ideas.

The Geneva conference brought very few ideas from each side that had not been independently discovered by the opposite team. In one of those rare cases, the Princeton Plasma Lab team, led by the then-still-young astrophysicist Lyman Spitzer, suggested an idea for a different kind of magnetic device for plasma confinement. As a representative of astronomy, Spitzer called it a stellarator. It was one of the ingenious geometries for magnetic fields capable of containing hot plasma.

Artsimovich had great difficulty in accepting that there was no counterpart to such an invention on the Russian side. In his review article for the conference, he subconsciously created the impression that something similar was in our portfolio, too. It was then that Leontovich stood up and essentially rejected the very notion that it might be so. His intervention was a remarkable lesson of scientific integrity—and, I would say, even human decency—for us young scientists. After all, at the end of the conference both teams agreed in a somewhat teasing way on the unofficial score of three to one in favor of the Americans in experimental plasma and fusion physics and one to one in theory. That made all of us who were Leontovich's pupils especially proud.

10 / Landau

➢

No matter how much we younger generation admired our own mentors at Kurchatov, Lev Landau as a physicist stood above them all. He simply reigned over Soviet theoretical physics. Even according to his own half-joking, half-serious table of ranks, he was number one, followed by the next few seats: empty, empty, empty. Only then would Landau recognize someone like Zeldovich, another great name in Soviet science. We younger guys teased the contenders that the most they could hope for was to be moved to sixth place instead of sixth and a half.

Throughout all of my Kurchatov Institute years, I was immensely fortunate to have the opportunity of attending the famous Landau seminar, held once a week at 11:00 A.M. on Thursdays at the Kapitsa Institute of Physical Problems. The room was usually filled with fifty or a hundred of the best theoretical brains of the country from many different institutions. Like me, most of the scientists had gone through the Landau minimum examination.

The seminar was an intellectual smorgasbord, the crème de la crème of theoretical physics. Admission was free—anyone could come and attend—but the criteria for being part of this elite was hidden in the style and the very spirit of the seminar. The ideas, the phrases, and the formulas on the blackboard were the ultimate expression, in condensed form, of the most sophisticated contemporary wisdom of our science.

Landau was a very sharp interlocutor. He did not lavish on his companions or colleagues a nonstop stream of bright stories but made his remarks very brief and straightforward. But all of the participants felt that they were intellectual diamonds. Even if he delivered a paper himself, it wasn't a very long talk. He tried to find a very adequate and brief description for everything. His own writings in physics were usually extremely condensed.

We young pupils who had passed through his selection criteria were obligated from time to time to report on recent articles in current scientific literature. He himself would go through the table of contents of the important magazines and leave marks on what he regarded as the most interesting titles. When the time came for each of us to present a paper, we went through the list marked by Landau and selected a topic. Then we would spend many hours preparing a final report, trying to understand the main conclusions of the authors. Then we reported our findings before an audience at the seminar. The day of presentation was awaited impatiently by everyone, not so much to learn what was new in the current article, but rather to find out Landau's final verdict.

His comments were always on target. Sometimes, full of irony, he would get up from his chair, take his chalk, and within a few seconds give a much more elegant and short derivation of the final formulas and results of the article. Often, he wouldn't even bother to do so, concluding that the article was absolutely irrelevant and that it didn't bring any new perspective to the subject. In such cases we reporters would feel especially vulnerable when he asked, "Why didn't you guys see from the very beginning that this is nonsense? Why did you take up our time?"

Occasionally, to preempt such a reaction from Landau, some reporters tried to be smart and prefaced their comments by saying, "I have to report to you on one rather silly article." However, Landau's reaction was unpredictable. He might easily interject, "Don't treat every author as a fool."

To our surprise, we discovered that there were a few theorists in the world whose rank Landau considered even higher than his own, and he always spoke about them with great respect.

Landau met most of these giants when as a young theorist he was sent by the Soviet Academy of Sciences to the best European centers of excellence in physics. The most important of all these places, he confessed later, was Bohr's institute in Copenhagen: "I consider the Danish physicist Niels Bohr my teacher. He taught me to understand the uncertainty principle of quantum mechanics."

His maturation under Bohr did not pass unnoticed in the Soviet environment, contaminated by a demagogy of quasi-scientists. In the eyes of these ideologues, Landau personified the most hated principle of contemporary physics: the principle of uncertainty. They considered Landau a walking disseminator of the pernicious, reactionary, idealistic philosophy of Einstein and Bohr. Despite these accusations, however, Landau was not afraid of such attacks. With a few of his colleagues back in Leningrad, and then from 1936 onward in Moscow, he bravely fought the war against the incompetent interpretation of contemporary physics by pseudophilosophers. Landau, Tamm, and a few other young physicists of the early thirties constituted the most outspoken group of young Turks defending the interests of true science.

Landau's main preoccupation was the quality of research. In one of his most eloquent deliberations in 1936, he bitterly attacked the current state of the art in that sphere. He wrote that even if the pool of nominal physicists in the country were rather modest, many of those nominally labeled "physicist" were simply worthless. Although he was a great supporter of improving educational standards, he recognized that science had its own logic.

Once he remarked that scientists belong to "class society": the division between classes is according to the scientists' actual objective scientific standing. Einstein, Bohr, and a few others belong to the highest level, the first class. The rest of them are distributed among the lower tiers. By definition, the admission to this hierarchical "class society" is granted to those who can relate themselves to every specific class, evaluating modestly and justly their own potential. As Arthur Conan Doyle once said, "Mediocrity knows nothing higher than itself, but talent instantly recognizes genius."

Landau was modest enough. He well identified a place for himself in the second class, below Einstein, Bohr, and a few others. However, in our eyes, he was almost a god, or at least a Mozart of theoretical physics. He developed bright ideas and formulas with gracious ease, we thought. For that reason we regarded every minute in his company as a celebration.

Outside of the regular seminars, we most appreciated informal conversations every Saturday morning at 10:00 A.M. Joining us in the large room assigned for his postgraduates, Landau comfortably occupied the chair and conducted a kind of general conversation about life, about everything. I felt that he enjoyed such deliberations and that in a friendly way he regarded himself as more than simply a teacher. He even asked a few questions, indicating his interest in our personal lives.

When I decided to marry, Landau reacted in quite a peculiar way. I was the speaker at the seminar, reporting on my own work on the theory of solitons.[1] However, before I was even able to open my mouth, Lan-

1. Nowadays the study of solitons is "à la mode" not only in plasma physics, but in literally every field within the hard sciences. But in the summer of 1959, it was one of the few pioneering works. I still feel proud of that paper and Landau's reaction, which was quite positive.

dau turned to the participants with a sort of wry smile and said, "You know, colleagues, Roald is a supporter of endogamy."

"Endogamy—what does it mean in that context?" several participants in the seminar demanded. He said, "I'm talking about his marriage inside theoretical physics!" Everyone laughed. That was his way of congratulating me. He meant that my father-in-law, Frank-Kamenetsky, was a theoretical physicist.

By the end of 1959, a year after the Geneva conference, though I was already established in the world of fusion science and plasma physics, I still had no official recognition. As it is done in our system, the doctoral degree is essentially introduced in two steps. The lowest grade is the candidate degree, which usually requires a few years of work and the production of an extended thesis. The Kurchatov Institute, despite unveiling the secrets of fusion research and opening its lab to foreign visitors, still did not have enough momentum for change inside its bureaucratic system. The only way to present a dissertation was to go through classified channels of evaluations and final hearings.

I thought it would be ridiculous to follow this path, since most of my results had been published in open scientific magazines and reported at international conferences. With some level of embarrassment and hesitation, I decided to bring this subject to Landau's attention. He was very forthcoming: "Of course I am ready to help you pass the presentation of your thesis at the scientific council of my institute, but final approval should come from Peter Leonidovich Kapitsa, the director and founder of this institute. I will talk to him."

A few days later, Landau brought back a positive message from Kapitsa. Before I could present my dissertation, as a prerequisite to final presentation I would have to deliver a substantial talk at Kapitsa's seminar.

Two weeks later, I was speaking in the same familiar room where the Landau seminars had taken place, but the audience this time was much broader—a mix of general physicists, including quite a few theorists. The majority in the room, however, were Kapitsa's pupils or colleagues in experimental physics.

Kapitsa was very active. He asked questions, trying to build for himself a clear physical picture of the processes I was talking about. Landau, to my surprise, was low-key. Apart from a few insignificant remarks, he said nothing during most of the seminar. Somehow I felt that Kapitsa and the participants of the seminar were pleased. When Kapitsa invited a small group of guests to join the traditional ritual tea party, I felt ashamed and embarrassed to be the center of such activity, creating a fuss out of a rather stupid formal procedure.

Kapitsa probably sensed my uneasiness and said, "Don't worry. All of us had to do it. After all, you should remember—love is passing, but the diamond ring remains forever." As Landau's pupils we knew, of course, that in the late thirties in the early stages of Kapitsa's and Landau's scientific careers, they had been engaged in very active creative cooperation.

Kapitsa had experimentally discovered that liquid helium cooled below a temperature of about 4 K (4 degrees Kelvin) acquires the extraordinary quality of superfluidity. Its flow through pipes is completely frictionless. Though this phenomenon was clearly a companion to the phenomenon of superconductivity, it was clear from the beginning that the explanation had to be very different.

The first theory of superfluidity was articulated by Landau in 1940 and by my time was included in all the textbooks. Kapitsa and Landau, working jointly on different but closely coupled aspects of this part of physics, represented one of the most outstanding examples of how an experimentalist can cooperate with a theorist. Soviet physics had to be proud of the contributions of these two outstanding men. Eventually, both men were awarded the Nobel Prize, though on different occasions.[2]

Presentation of my candidate thesis went smoothly, except for perhaps a panic that ensued just a few days prior to the event. The bureaucratic office that had to issue the final okay for declassification of my manuscript created an unexpected obstacle. In order to have a declassified presentation, the announcement of the presentation had to be published in an open newspaper. But to have the right to contact such an open publication required that I prove that the dissertation was declassified and contained only open materials. A few days prior to the meeting at the Kapitsa Institute of Physical Problems, I discovered that to collect all of

2. As a matter of fact, Landau was decorated a few years prior to Kapitsa. But that does not necessarily indicate that the theoretical explanation was more important than the first experimental discovery.

these bureaucratic papers with bunches of signatures and stamps required at least a few months.

In a state close to desperation, I phoned Kurchatov's secretary. She suggested that I come to the office of the director and she would find a way to get me in touch with Kurchatov. Thus, half an hour later, she opened the door and pushed me inside Kurchatov's office and in front of Kurchatov and Yefremov, his supervisor from the government.[3]

I thought my mission was doomed, but even before I could open my mouth, Kurchatov, who was in a very good mood, introduced me to Yefremov as one of his young geniuses and asked how he could help me with my problems. A few minutes later, I left with a paper carrying Kurchatov's signature. This one signature was worth everything, and the editor of the newspaper *Evening Moscow* immediately accepted my announcement for publication.

The final outcome of the official presentation ceremony was predetermined, as Landau predicted. He himself said a few nice words. In conclusion, Kapitsa gave strong support, even saying that with a few more formalities resolved in advance, the presentation could spare me time and liberate me from the next step of moving to a doctoral degree. I didn't know yet that my cooperation with Landau, and the chance to benefit from his reign in Soviet physics, was almost over.

One year later, in 1961, the brain of Landau, the Mozart of theoretical physics, was lost to science forever. It did not happen because of the envy of a scientific Salieri, though the system provided fertile soil for the mass production of such men. The end came as a result of a bizarre car accident. No passenger in the car was hurt except Landau, who was gravely injured—so injured, in fact, that he went through several clinical deaths and successive reanimations before his situation stabilized.

The best international team of brain surgeons came to Moscow to perform a complicated operation, but all of the power of modern medicine was helpless. Though he recovered physically, Landau was never able to return to his beloved science. Something extremely elusive was lost from his brainpower. During his remaining eight years, he never attempted to recuperate his genius for the most elegant formulas and

3. Yefremov was the same man, in fact, who had advised delegates to the Geneva conference not to tell the world more than a kopeck's worth of our scientific knowledge.

derivations. Moreover, he never again attempted to be engaged in any serious discussions on science.

A few years later, while he was still alive, I remember him sitting on the sofa in his old apartment, showing off a cartoon presented to him on his fiftieth birthday. The artist had depicted the better times: Lev Landau as a lion, king of the beasts, surrounded by his pupils. (*Lev* in Russian means "lion.") Landau commented with a sad smile, "Pitiful lion! That's what I am now."

For many of us students and admirers of Landau, it was unimaginable that such a mind could not function anymore in physics. There was even a theory that maybe Landau, unwilling to go down in his own table of ranks, deliberately refused to do science anymore. Nevertheless, it was most painful for all of us to talk to him—just as if nothing had happened—but at the same time to carefully avoid any discussions of science and theoretical physics.

With sadness I recall his advice to us, his pupils: "Life is short. Don't try to take a job you will be unable to accomplish. Never try to make only great discoveries; otherwise, you may be victimized by your ambitions."

He himself followed these rules throughout his scientific career. His genius was able to bring mathematical elegance and beauty to the physical argument—even to subjects that might have been considered rather narrow and, by definition, below the merits of grand scientific awards.

11 / Kapitsa: The Lone Warrior

➢

When I first met Peter Kapitsa, he was already in his mid-sixties. At that time, in the late fifties and early sixties, he had already outlived most of his contemporaries, and one could easily see in him a deliberate detachment from the mundane everyday bustle, such as petty budget wars for science or competition to fill the vacant chairs in different departments of the Academy of Sciences.

His long-standing independence was always the underlying feature of his character. In a rather strange way, but in one clearly consistent with such behavior, he never gave outsiders—or even friends—a chance to assess who he actually was. Only at the end of his life did he invite associates and colleagues to read letters and examine the records and documents he had carefully kept and accumulated throughout his life. So who was he, to survive and to reach such a unique and outstanding position in twentieth-century Soviet science?

Young Kapitsa had already made up his mind to become a physicist when the revolution of 1917 broke out. He was ready to resist any attempts to become involved in the whirlwind of events. He and his contemporary, Nikolai Semenov, exchanged an oath with each other to dedicate their lives to serving science. They were quite ambitious and had no doubt that eventually both would become famous and rich.

However, food shortages and the hunger of the civil war period did not give them the chance to establish a comfortable exile from politics. Deliberate detachment from harsh realities was impossible. There was not much interest around to support the future pride of Soviet science, and Kapitsa did not escape his own cup of suffering and personal tragedy. In the famine and the epidemics, he lost his wife and both of his children. His colleagues knew that because of this tragedy he was close to suicide. But in 1921 he got the chance to get out of the vicious cycle of life and go to the United Kingdom on a fellowship. Ernest Rutherford, one of the greatest atomic and nuclear physicists, became his mentor at

Cambridge in the famous Cavendish Laboratory. The atmosphere of creative physics became his aqua vitae, reanimating Kapitsa's soul.

Rutherford discovered that this enigmatic Russian happily combined scientific genius and engineering creativity. He was eventually promoted to the head of the lab as deputy responsible for bringing new, and what are now called exotic, technologies to the arsenal of nuclear physics. Kapitsa took a patent on his technical inventions that generated the highest possible magnetic fields. In this, he achieved record values of 500,000 gauss. His contemporaries could not imagine this, even in their wildest dreams. He thought that putting matter under extreme conditions—a strong magnetic field—would reveal an interesting new feature.

The attainment of extremely low temperatures was another of Kapitsa's passions. He designed several machines to compress and liquefy different substances to attain the lowest possible temperature. The originality of Kapitsa's approach to research in physics soon brought recognition from officials, and the Royal Society raised money to build a special experimental lab, the Mond Laboratory, to implement the ideas of their Russian visitor.

In this environment, life started anew for Kapitsa, and he eventually married again. His wife, Anna, provided not only the support of a kindred spirit, but family ties to the great figures of past Russian science. She was the daughter of Alexei Krylov, an outstanding naval engineer who rebuilt the Russian navy after it was defeated in the Russian-Japanese War of 1904. This genetic family line was also related to Mendeleev, author of the periodic table of the elements. Ties with the past and a lifeline to future Soviet science are what the Kapitsa family provided for the young Soviet republic.

However, the rulers in Moscow had a different perception of the situation. On one of the Kapitsa family's regular visits to Leningrad for summer vacation, the government informed Kapitsa that his exit visa had been canceled and he would not be allowed to do any more work in foreign scientific institutions.[1] Kapitsa never saw Rutherford again.

The Soviets indicated that they would be willing to give Kapitsa a

1. There are varying interpretations of the reason for this detention. Some believe that the regime denied him the right to travel because they wanted the services of this outstanding scientist to be put toward the development of the Soviet Union. Others believe that Kapitsa was detained in retaliation for the defection of another famous scientist, George Gamov, who later became one of America's leading experts in astrophysics.

budget to start a new institute in Moscow, and even to negotiate with the British authorities to bring back some of the lab equipment he had accumulated in Cambridge.

On returning to Moscow in 1935, Kapitsa had to start everything from scratch. He had to establish contacts with Soviet officials and reestablish friendships with his colleagues. In the atmosphere of the country, there were already signs of the approaching terror of the mid-thirties. In fact, Sergei Kirov, second man in the party and hero of the masses, had been assassinated in Leningrad just a few months after Kapitsa was detained. All of this changed the spiritual atmosphere and mood of the country. Given the mounting tensions, fellow scientists avoided Kapitsa as if he might be someone who carried the contagious disease of the bourgeois West. Officials quickly taught Kapitsa a lesson about the omnipotence of the newly born Soviet bureaucracy.

The only kindred spirit Kapitsa found in that society was the ailing physiologist Ivan Pavlov, who would say, in conversation with his younger colleague, "Here I am the only one to speak what I think. After I die, you ought to do the same thing. It is so important for this country."[2] Kapitsa was so emotionally touched by Pavlov's confidence in him that for a while he considered making a transition from physics to biophysics and physiology by joining Pavlov's own institute.

Fortunately for physics, Kapitsa never made this transition. But he was a faithful heir to Pavlov's civil will to continue to raise the voice of integrity, the voice of dissent.

In a confidential letter to his wife in Cambridge a few months after he was ordered to stay in Russia, he wrote: "Bureaucracy strangles everyone. However, despite my criticism, I believe the country will be able to resolve all of its problems victoriously. I believe that one could prove not only that the socialist economy is the most rational, but that it would create a state system, meeting the highest demands of the human spirit and ethics."

However, Kapitsa recognized that between this long-range optimistic forecast and the current state of affairs, especially in starting a new scientific enterprise, there was an unsurmountable conflict: the govern-

2. Kapitsa in a letter to his wife. This and all subsequent references to Kapitsa's correspondence are from Nauka Publishing Haus, P. L. Kapitsa, *Letters on Science* (Moscow: 1989, in Russian).

ment. "Probably the only resolution is to look for the exclusive situation, so to say, the one under direct supervision of authorities like plants in a greenhouse."

He patiently started the long, systematic siege of the corridors of power. He tried to see bosses, to whom he talked about science in general and his own ideas on how to build a new institution. He wrote letters to second-ranking functionaries and heads of different offices in the government and party, eventually reaching the highest echelons of power. He was not humiliated even by negative answers.

Slowly, step by step, Kapitsa overcame one obstacle after another, forcing the government to accept his conditions. Only a few months after a bitter letter to Molotov, he wrote to his mentor and friend Rutherford:

> My dear professor:
>
> Life is an incomprehensible thing. We encounter the difficulties even when we are trying to study certain physical phenomena. So I think a human being would never be able to understand human destiny, especially one as complicated as mine. After all, all of us are only tiny particles carried by a stream which we call fate. The only thing which we can do is slightly navigate ourselves and try to stay on the surface. The stream is carrying us. In any case, the country seriously contemplates and hopes that science can develop and occupy an important position in the social structure.

By mid-1936 Kapitsa's new institute on the Moscow River was close to completion. After long negotiations, the hardware and laboratory equipment in Cambridge were brought to Moscow by the Soviet government. After carefully screening the candidates, Kapitsa started to build his own team of scientists. With great pride, he wrote an account to Rutherford and added the bad news: "I still feel like a half-prisoner because I have no chance to travel abroad, to see the world, to visit labs. It is a great loss. Undoubtedly, it will lead in the final account to the narrowing of my expertise and ability."

However, on the internal front, things were moving much faster. He created a new intellectual environment in his own institute with the brightest young theorists like Landau joining him. Moreover, there was another, no less visible, achievement in his tireless battle with the apparatchiks. They were not invincible. He could see the results of his persistent attacks. The apparatchiks were unable to ignore his critical remarks, his defense of the values represented by the scientific community. He decided to further intensify the campaign. It was too early to say a farewell to arms.

Meanwhile, the country was approaching the year of the Great Terror, 1937. The scientific community was not missed in its wake. The situation had advanced beyond the simple interruption of indispensable scientific ties with international, mostly Western, science. It went much further. What was at stake was an issue of life and death. Many scientists were arrested during the great purges, most of whom never came back from the camps.

Kapitsa, still a true believer in socialist ideals, was concerned that such a thing could annihilate the country's scientific potential so badly needed for the socialist construction. He was worried that there was a massive campaign to arrest theoretical physicists. So many of them were detained that some courses in the Moscow State University curriculum were left without lecturers. Kapitsa's dismay culminated in the arrest of one of the brightest physicists, Vladimir Fok. Vowing to hesitate no longer, Kapitsa addressed his appeal directly to Stalin.

"Fok's arrest is nothing more than an act of rough treatment of scientists, which creates, as in the case of rough treatment of machinery, nothing but damage," he wrote to Stalin. "This inflicts damage on all international science. Such treatment of Fok creates an internal reaction here and among Western scientists similar to the one created when Einstein was expelled from Germany."

In a more detailed letter to the deputy to Prime Minister Molotov, he brought an additional argument in favor of Fok: "This is a man absolutely isolated from life because of his almost complete deafness. I cannot imagine such a man committing a substantial number of crimes. It seems to me absolutely improbable that Fok intentionally suggested a wrong theory. After all, it would be too easy to disclose; but it is even more important that Fok is too prominent a scientist to do such a thing. You see, it is like being a great musician—for him to play falsely would mean to torture his own ear."

Kapitsa's appeal created a miracle. The hand of the executioner was stopped. Fok was released and saved for international science, which eventually benefited from his many articles.

For Kapitsa himself, Fok's miraculous rescue was a very strong signal that Stalin had heard his voice. It was experimental proof that his carefully designed scenario—to acquire the special status of "most favored national scientist"—was working. It gave him a strong boost to continue his letter writing. Over a period of more than a decade, Kapitsa established a unique hotline of communication with one of the most frightening dictators of the century, Stalin. Until Kapitsa's death, Stalin became the principal recipient of Kapitsa's deliberations on the role of science in Socialist Russia, or requests for certain types of promotions or help to facilitate scientific projects of his own or of his colleagues.

Kapitsa launched a courageous criticism of the Soviet bureaucracy, including the closest aides and subordinates of Stalin himself. No one since the time of Julius Martov and Lev Trotsky had been brave enough to touch Stalin, even in the slightest negative way. Kapitsa's unprecedented approach to Stalin was a feat in itself. Kapitsa knew that what was happening was not the monologue of a lone speaker. It was dialogue, even if Stalin did not respond to his letters. There were always indications that he had been heard. Furthermore, the ultimate proof of Stalin's appreciation of and interest in Kapitsa's messages was the very simple fact that Kapitsa was still there, untouched and able to continue writing the letters.

It cannot be denied that to a certain degree Stalin was amused and entertained by this person who was so unlike the sycophants that surrounded the Father of the Nation. On the other hand, such a form of correspondence, which was obviously kept confidential, might also have awakened Stalin's nostalgic, sentimental feelings for the early illegal, conspiratorial letters that the Bolsheviks had written to each other before the October Revolution of 1917.

Whatever the reason, Kapitsa soon found himself playing negotiator and moderator, trying to pacify the chief bandit.

Kapitsa's pen never trembled. He was always rational and sober.

In April 1938, he wrote:

> Comrade Stalin,
>
> This morning the scientific collaborator of the institute, L. D. Landau, was arrested. Despite the fact that he is only twenty-nine years old, he, together with Fok, is the most prominent theoretical physicist here in the union.

A year later, desperate but persistent, he wrote another letter:

> Comrade Molotov,
>
> While recently working with helium near absolute zero, I discovered the need for a theory of the new phenomenon that will clarify one of the

> most mysterious fields of contemporary physics. I need the help of a theorist here in the Soviet Union. This field of theory, which is important to me, is a field of great competence of Landau. However, the trouble is, he has been under arrest for a year already.

Kapitsa's promise was fulfilled. Soon after his release from the Lubianka,[3] Landau came up with the theoretical explanation of this mystery of contemporary physics, the superfluidity of liquid helium.

It's remarkable what Kapitsa, a lone warrior, achieved in his uncompromising battle against the regime. However, I believe it was a well-thought-out, deliberate decision to count only on his own strengths and to assume sole responsibility for his letters.

Apart from his courageous stands against governmental repression, Kapitsa was also able to force the government to work at his request to accelerate the construction of the new buildings of the institute and to keep a supply of hardware and components that were critical for the scientific work of the institute—in a time of great shortage in the country.

In the end, perhaps it was not Kapitsa who was kept by the government as its hostage, but the recipients of his letters who became Kapitsa's hostages. After all, when it was necessary, he fought them using their own weapons—by passionately talking about the superiority of the socialist system.

3. KGB headquarters.

12 / A Hostage to Politics

➢

Despite his intense years of conflict and battle with the authorities, Kapitsa's travails did not seem to affect his genuine belief in the ideas of communism. Otherwise, it would be difficult to explain why he tried so hard to invite the best European scientists, including Nazi refugees, to the Soviet Union.

In October 1943, Kapitsa wrote to Molotov:

> Esteemed Vyacheslav Mikhailovich:
>
> Today I have learned that the Danish physicist Niels Bohr flew to Sweden. He is the most prominent scientist, founder of contemporary atomic science, and a Nobel Prize winner. Bohr respects the Soviet Union. I believe it would be very good and sensible if we offered Bohr and his family hospitality.[1]

Two weeks later, Kapitsa wrote a personal letter to Bohr conveying the invitation. This letter was delivered to England after Bohr returned from the United States, where he had already been briefed on the atomic bomb. It is not surprising that the letter from the USSR, delivered through diplomatic channels, caused real alarm in London. Just as Kapitsa had to coordinate his original letter to Bohr with the Soviet officials, the reply from England had to be approved by the British intelligence service. Needless to say, Bohr didn't come to the Soviet Union.[2]

1. This and all subsequent references to Kapitsa's correspondence are from P. L. Kapitsa, *Letters on Science* (Moscow: Nauka Publishing Haus, 1989, in Russian).

2. As one of the first atomic scientists deeply concerned with the threats of the potential nuclear arms race, Bohr asked for an audience with Prime Minister Winston Churchill. Bohr surprised Churchill with his approach to the whole problem. Bohr's position was that the secret of the atomic bomb would eventually be rediscovered by the Russians. This is why, to avoid bad feelings and future negative impact, it would be sensible and wise to share the secrets generously with the Soviets. History preserves the memory of how two great people felt after such an encounter. Bohr was extremely embarrassed by his inability to deliver his message in a way that would be understandable

The failure to bring the patriarchs of Western science to the Soviet Union did not stop Kapitsa. He designed another extraordinary scenario: to bring back to the Soviet Union the best Russian thinkers who had gone to the West after October 1917. He thought, at the end of World War II, that the time was ripe for such an attempt. The world was impressed with the heroism of the Soviet army and the Soviet people, and Russia was evidently regaining international respect and sympathy.

Kapitsa approached Stalin, according to what he told me much later, with an idea to start a confidential correspondence with outstanding Russian émigrés living overseas. The success of such a plan clearly would have enhanced Stalin's efforts to gain the technological and scientific edge over the West. The postrevolutionary wave of Russian emigration to the West carried with it hundreds of bright scientists and engineers. Quite a few of them established themselves in the United States as leaders in their fields of science and engineering, and some of them even as prominent capitalist entrepreneurs.[3]

How naive Kapitsa must have been, hoping to bring such people, and many others, back to Russia to work for Stalin and Beria!

Of course, during this time, Kapitsa was receiving wide recognition from the regime for his scientific work. He received awards and decorations, commemorating his achievements in science and technology. Kapitsa himself was honored with the highest degree of recognition on May Day 1945. Stalin signed the decree of the presidium of the Supreme Soviet, declaring Kapitsa the Hero of Socialist Labor. His Institute of Physical Problems was awarded the Order of Labor Red Banner.

However, soon Kapitsa's letters to Stalin acquired a new and suddenly alarming pitch. He complained about Comrade Beria:

> Now getting in touch with Beria in conjunction with our special committee [that very highest body to supervise the atomic program], I have gotten a particularly clear feeling of how impermissible his [Beria's]

---

to Churchill. The latter was so irritated that as physicists tell the story, after Bohr left, he asked rhetorically, "Why isn't this gentleman in prison?"

3. Igor Sikorsky, one of the great names in aviation, was one, as well as George Kistiakowsky, a talented chemist, who became an important part of the Manhattan project and later Eisenhower's scientific advisor. There were Russian names in almost every branch of science. Many of them were outstanding. Stepan Timoshenko, whose theory of elasticity is still used by material scientists and engineers; and George Gamov, in nuclear astrophysics, who developed a model of the universe after the big bang theory as a giant exploding nuclear reactor. His theory is the bible of contemporary cosmology.

attitude is toward the scientists. . . . When he wanted to appoint me, he simply ordered his secretaries to call me up. When Witte,[4] the minister of finance, wanted to see Mendeleev, he himself paid a visit.

The letter ends unexpectedly:

It is time for comrades of Beria's type to start learning how to respect scientists. All of this is forcing me to realize that our country is not yet ready for intimate and fruitful interaction of political forces with the scientists.

Kapitsa realized that he was absolutely incompatible with Beria and asked Stalin to agree to his resignation from the special committee of the atomic program.

The conflict with Beria escalated very quickly. Soon Kapitsa discovered that he was engaged in a mortal duel with one of the biggest executioners of the century. He saw that his only chance of survival would be to carry on this dangerous battle before the eyes of Stalin, the principal referee. Like David against Goliath, he wrote an almost suicidal letter in November 1945:

Comrade Stalin:

It has been almost four months since I started attending the meetings and actively participating in the work of the special committee. In the organization of the work for the atomic bomb, I believe there are a lot of abnormal things. It is essential to have more trust between scientists and statesmen. In our country, we do not educate people to have enough respect for scientists and science. Comrades Beria, Malenkov, and Vosnesensky[5] behave in the special committee as supermen, particularly Comrade Beria. True, he has the conductor's baton in his hand. It would not be so bad, but the next after him, the first violin, should be taken by a scientist because the violin gives tone to the whole orchestra. But the main weakness of Comrade Beria is in the fact that, as a conductor, he has the duty not only to wave his baton, but to understand the musical score. Unfortunately, Beria does not.

So, I told him directly: "You do not understand physicists. You need to ask the scientists to make judgments on these issues." His answer to me was that I do not understand human beings at all. I suggested to him that I am ready to deliver introductory courses in physics: "Come to me at the institute. After all, it is not essential to be a painter in order to understand the paintings of others."

4. Sergei Witte. Kapitsa was comparing the way the czarist government had treated the country's scientists in contrast to the Stalin regime.

5. Nikolai Vosnesensky.

But I failed with Beria. I don't like his attitude toward scientists. For example, he wanted to see me a couple of weeks ago. He made an appointment, invited me for a certain day and hour. But every time he canceled the appointment. Probably, he was doing it intentionally to humiliate me.

The dramatic letter ended:

I wish Comrade Beria would see this letter, because it is not a denunciation but useful criticism. I would be ready to tell him myself, but it is very difficult to see him.

Kapitsa felt like a mouse cornered by a dangerous and wicked cat. And he found what was probably the only solution: appealing directly to the protection of the chief cat.

A few months later, Stalin hinted that a kind of safeguard could be granted. It was the only letter Stalin ever sent back to Kapitsa during Kapitsa's seventeen years of writing to Stalin:

Comrade Kapitsa:

I have received all of your letters. They are very instructive. I hope one day we can meet and talk about them.

But what would be the price of protection? Beria was systematically building webs of intrigue against Kapitsa. He used what he thought was a weak point of the great scientist: Kapitsa's obsession with inventions. After all, Kapitsa was not alone in that passion. Even Einstein's secret life was to invent and apply for patents. As a matter of fact, both scientists were inventing in almost one and the same area. While Einstein was captivated by the technology of refrigeration, Kapitsa was promoting in Soviet Russia his ideas on liquefying oxygen and other gases relevant to the Industrial Revolution. While Einstein spent a lot of his time defending his patents in legal battles, Kapitsa delivered his technical genius for the good of the Socialist homeland in a different way. To facilitate and accelerate the technology transfer from his brain to operational industrial plants, he agreed, for the time being, to function as president of a special consortium known as Chief Oxygen.

As a crafty psychologist, Beria recognized that this second side of Kapitsa might be much more vulnerable than the research side of his work. That was the battlefield where Beria launched his offensive. Kapitsa was accused of not fulfilling his socialist commitment to launch planned installations of Chief Oxygen. Without difficulty, Beria recruited a number of "renowned experts" who were ready to testify against Kapitsa. Kapitsa wrote a desperate letter to Stalin telling him that he felt like Othello surrounded by Iagos: "It is a Shakespearean tragedy; and what will be the outcome? That depends on you."

The final verdict was signed by Stalin on August 17, 1946. Kapitsa was removed from all of his posts, not only from Chief Oxygen, but also from the directorship of his beloved Institute of Physical Problems. Many years later a witness disclosed a conversation he had overheard and later reported to Kapitsa. Stalin had said to Beria: "I will fire him for you, but you do not touch him."

Kapitsa was sent to his *izba* (country house), where he remained under de facto house arrest. Without his pupils and assistants, without sophisticated hardware and instruments, he spent almost ten years of his life in solitary seclusion. Despite this, no one, not even Stalin or Beria, was able to separate him from physics. He converted his wooden *izba* into an improvised physical laboratory, and while his colleagues and friends throughout the world were moving to larger and larger super-synchrotrons and other sophisticated installations of modern physics, he undertook an almost hopeless approach to "physics at home." With the most modest means, he performed studies of hydrodynamic flow in films and discovered capillary wave generation. It was an elegant piece of science akin to water drops sliding on window glass. Kapitsa was living proof that genuine scientific issues can never be exhausted.

I think Kapitsa took special pleasure in sending the reprints of several papers on his homemade physics to Stalin only two years after the beginning of his seclusion. The first page of one reprint contained a handwritten note that read: "One is not a scientist who simply carries on scientific research, but a scientist is one who cannot *not* carry on scientific research."

In the meantime, Kapitsa's mind had not been preoccupied solely with available-at-home experiments. Removed from the atomic bomb program, Kapitsa moved to the opposite extreme. He contemplated a defense against atomic bombs. Eureka! A solution could be found. He developed the concept of an absolutely exotic technology—high-intensity energy beams—which could be formed by extremely powerful microwave emissions.

In the summer of 1950, Kapitsa wrote a personal letter describing his masterful idea to Georgy Malenkov, the closest associate and most probable potential heir to Stalin. It could now be described as an early precursor to SDI (Strategic Defense Initiative). In a subsequent phone call, Malenkov suggested to Kapitsa that he write a more detailed letter to Stalin. In response, Kapitsa expressed doubt that Stalin would read it. Malenkov assured him: "Peter Leonidovich, Comrade Stalin is reading not only the letters you write to him directly, but also those that you are writing to me."

The hotline between Kapitsa's *izba*-under-siege and the Kremlin was functioning. However, the ailing Stalin was unable to undertake quick action to restore Kapitsa to full rights and to give him a chance to work under normal conditions. After all, Beria was still a man of significant influence. As it turned out, Beria outlived Stalin by a few months.

It was not until Beria was arrested and "put on trial" that Kapitsa was brought back from disgrace. The *izba* itself was renamed the Physical Laboratory of the Academy of Sciences, and Kapitsa was put on the payroll with a salary of 6,000 rubles a month—at that time, the salary of a typical Soviet professor.

Kapitsa's comeback into a new political environment with Khrushchev as leader of the state took place soon after the meeting between the two great men at the end of 1954.

"How much time would you need to develop your high-powered microwave devices?" asked Khrushchev.

"About five years," was the reply.

"That's too long."

Kapitsa was again in the office of the director of the institute, sending messages on the hotline to recipients who changed only at the Kremlin end.

It was after Khrushchev's trip to England and Kurchatov's speech at Harwell that Kapitsa learned about the new ideas in nuclear science to develop controlled thermonuclear fusion.

"That's it! This is the area—the problem that could be handled with the use of powerful microwave emissions," Kapitsa most probably said to himself. He was not about to adopt the way suggested by Sakharov and Tamm, based on confinement of hot plasma within a magnetic field. At that time, very few people—only Kapitsa's closest friends and colleagues, including academician Fok and Kapitsa's own son Sergei—had access to his secret ideas on microwave-generated fusion. Most probably, even Landau knew nothing about it when he introduced me to Kapitsa before the session at which I presented my candidate dissertation. The special interest and warm reception for my talk at Kapitsa's seminar on plasma instabilities might have strengthened Kapitsa's belief in the uniqueness of his own approach and underscored his skepticism about the wisdom of conventional controlled fusion.

However, I learned about Kapitsa's own theory only some years later, when Landau was already out of science and Kapitsa had decided to present his ideas for controlled fusion to the scientific community. A thick collection of manuscripts was sent to the government and to the presidium of the Academy of Sciences for consideration and adoption as a national program. The authorities decided to set up a special commission to assess the importance of the proposal and to come up with recommendations.

Among the members of this commission were Mikhail Leontovich and myself. The essence of Kapitsa's idea was to apply a concentrated microwave emission to the center of a vessel filled with heavy hydrogen, deuterium. The subsequent heating (analogous to that created by a very powerful microwave oven) of the gas would ionize it, he thought, and create a plasma that would eventually become hotter and hotter. With the power of the microwave emission above a certain threshold, Kapitsa sought to ignite the thermonuclear reaction.

The first reading of the materials disappointed us. The microwaves absorbed by plasma could in fact deliver a thermal energy proportional to the power of the microwave circuit. However, to accumulate and keep this energy so plasma could reach a high temperature required a special type of thermal insulation; otherwise, the whole process would be like putting water in a leaky bucket. Heat losses in such a case were not so difficult to calculate. That is what I did, together with Leontovich.

Leontovich quickly lost not only interest, but even the patience to deal further with Kapitsa's proposal. He asked me to stay there alone to argue with Kapitsa. In a huge room in an isolated part of the institute, Kapitsa had built several installations—of a medium size, compared to those at the Kurchatov Institute—to experimentally confirm his original ideas.

While it would usually be very difficult to get direct data on the temperature deep inside the hostile environment of hot plasma, Kapitsa based his evidence on indirect indications and proofs. I spent quite a few days watching what he was doing, studying the records, and talking to him.

He had a bizarre hypothesis, based on an assumption that hot plasma would spontaneously build a self-defensive thermal insulation shield to keep the heat inside. It sounded rather ridiculous to me, and I suggested to him that he find a theoretical collaborator. "Why don't you ask the theoretical physicists to make a straightforward calculation to justify such a model?" I said. "They would confirm or reject your explanation."

He answered, "That would take a much longer time."

"Now, Peter Leonidovich, I am ready to bet it could be done within only a few days. If you reject my assistance, why don't you ask some of your junior collaborators, your theorists? You have plenty of them in the Landau department. They could do it with great ease."

His answer really surprised me: "You know, in that way, I would deprive myself of getting the satisfaction, the joy of discovery."

I could not then, or ever, accept such a stand. Science is a collective undertaking, and teamwork is what makes science even more exciting. The sad irony of Kapitsa's life was that this lone warrior became a prisoner, a hostage of his loneliness, in everything—and most importantly, in doing the science of physics.

I felt very sorry for him and tried once more.

"Okay, if you don't want to invite the assistants or theorists, why don't you try direct measurements of X-ray radiation? This would undeniably deliver the message about the high temperature in the center of the plasma, even through the thick atmosphere of the cold and dense gas outside." By special arrangement, a student of mine, an experimentalist knowledgeable in precisely this type of measurement, was invited and asked to bring a special sensor. This device would be considered, from today's perspective, an early precursor to modern X-ray telescopes on space platforms.

Soon it became clear that the temperature was nowhere near the hundreds of millions of degrees Kapitsa had claimed. Kapitsa retreated but did not give up. He agreed to de-escalate his claim. The next modification of the measurements was prepared—to look for X rays associated with temperatures at least five to ten times lower. Again, no sign of X-ray quanta. Kapitsa's self-pride was in great danger. It was painful to witness the weakness of this great man, who chose to terminate the experiments. Alas, he preferred self-deception.

The final conclusion of our commission let Kapitsa publish his report in open scientific journals. There was a silent convention to let the old man continue his exercise, his obsession, at his own risk. The papers were sent to scientific journals, and some of them were finally published. The response of the international scientific community was polite silence.

Kapitsa continued his unrelenting pursuit of this idea for the rest of his life, which lasted for many more years. His son, Sergei, my very old and good friend, originally assisted him in most of these lonely undertakings, but later deserted the activity. Sergei later confessed to me that he had finally given up all hope of changing his father's mind.

Many years later, in 1984, when the country was approaching the advent of the Gorbachev era and perestroika, Kapitsa died, falling short of reaching his ninetieth birthday. Despite the deep respect the institute had for its founder, in its first move after his death it decided to immediately dismantle all of the alchemical microwave devices Kapitsa had accumulated. That was the end of the great obsession.

One can only imagine how much more tragic it would have been had Kapitsa directed his passionate drive toward an antiballistic missile (ABM) system, his early scenario for SDI based on microwave emissions. Gradually, Kapitsa lost interest in pursuing this idea. Years later, in the early seventies, I heard the rumor that the Ministry of Radio Industry, a prominent part of the Soviet military-industrial complex, was organizing an expensive research and development effort on microwave beams as an antiballistic kill weapon. But I never heard whether Kapitsa had had any part in this particular story.

The program eventually died—after wasting, I would assume, at least several hundred million rubles. It died not because of strategic reconsideration of the ABM concept, or even as a result of the ABM treaty signed in 1972, but because the hopes and interests of the military-industrial complex had shifted to different ideas and technologies, such as laser beams, which were thought to be much more promising.

13 / Fusion

➢

"Perhaps if we learned more about plasma physics, we could suggest a more sophisticated and intelligent approach. So wait for a while." That was the official message of the fusion team to Kurchatov.

But waiting was not something that Kurchatov did very well. I'm sure he considered Artsimovich and Leontovich overly pessimistic, lacking both the enthusiasm essential to ignite the spark and fervor critical for the success of the controlled fusion team's creative work. Perhaps this is why Kurchatov tried to circumvent the older scientists by reaching down directly to the younger members of the team. It was in such an atmosphere that he invited me for a second, very long, conversation in the summer of 1959.

When I briefed Kurchatov on the development of the theory of plasma instabilities, he nagged me, comparing all of my current statements with those that I had made about a year before at our earlier meeting, the notes of which he had kept in a special file book. I thought his notebook represented a remarkable example of self-discipline and self-organization.[1]

To Kurchatov's great disappointment, I told him about a newly discovered instability that was just coming from the still-warm formulas found by a very close colleague of mine, Leonid Rudakov, and me. We came up with theoretical predictions about a peculiar type of plasma motion, slightly similar to cyclonic front buildup in the atmosphere that appears as a precursor of bad weather. I told Kurchatov that these disturbances, called drift waves, were essentially developing into almost incurable instabilities. The irony is that we did not know at that time that the word *almost* would be removed as a result of three more decades of fusion research in many plasma labs around the world.[2]

1. He followed the famous old Russian saying "Trust but verify," a statement that many years later was so much liked by Ronald Reagan.

2. It is incurable and we have to live with it, building bigger and more sophisticated machines where such an instability, even if not removed completely, will let plasma maintain thermal energy long enough.

But Kurchatov was unable to keep his impatience under control.

"You skeptics and overcautious people!" he told me. "How many more people, how many more resources are needed to tackle this instability in the plasma? Could you suggest any experimental device that would provide the test bed for the instabilities predicted by your theories?"

I thought that the conclusions we were developing in the lab were not ripe enough to become a starting point for the implementation of an experimental program or any other larger-scale activity.

I replied to Kurchatov, "Do you think that nine pregnant women would be able to do in one month a job that takes nine months?"

He laughed and responded, "Okay, go and do your job. But keep me in touch regarding all important events."

When I reported this visit to Leontovich, he smiled slyly and said, "You were too modest. I know what we should ask from him. I'm going to make a phone call suggesting that he invest a little bit to move you from a dormitory to more decent living conditions. Maybe he could give you a room in a communal apartment."

Indeed, Leontovich's intervention did help. I was given a small room in a two-family apartment not long after my visit. Frankly speaking, even if this input from Kurchatov had not directly given a great boost to my fusion research, the apartment came at the right moment. I had just married, and my wife, Tema, and I were moving from one rented room to another. Without an official lodging in a communal apartment, we would have encountered real difficulties nine months later when our son, Igor, was born.

I have often recalled Kurchatov and his generous intervention. But not long after he secured our communal apartment, he was gone. In early winter 1960, he suddenly collapsed after a stroke, only a few days prior to the day my son was born. We named him in Kurchatov's memory.

Kurchatov's intentions to give a huge boost, a new impetus, to the troubled fusion program did not materialize. Perhaps if it had, it would have been premature. The development of a revolution in science is controlled by its own internal logic. To build a new, revolutionary concept, to make a breakthrough, requires a certain hidden incubation period, during which time there is the accumulation of experimental data, the painful assessment of difficulties, careful invention, and then the injection of new scenarios and explanations. Science and physics have always progressed in this way.

Somehow we, the nuclear physicists of the twentieth century, were spoiled by quick successes like the Manhattan project and, a few years later, a parallel breakthrough with a nuclear bomb on the Soviet side. Many of us, even such wise and experienced leaders as Kurchatov, thought that if an appropriate budget were given it would almost guarantee immediate technical progress in resolving any problems nature presented us.

If we physicists had subconsciously become somewhat arrogant, our punishment was not long in coming. Controlled thermonuclear fusion, unlike the uncontrolled one with its apocalyptic hydrogen bomb explosion, was not an easy nut to crack.

As it turned out, the very nature of plasma—the hottest state of matter and at the same time the least controllable substance—destroyed the legend of Almighty Science versus Nature. What was originally perceived as a quick victory over the new, inexhaustible source of energy became a long, protracted war against plasma instabilities. It created almost a deadlock and proved to be a setback and a warning of even bigger future failures of humankind, which saw itself as the master of nature.

The ultimate type of incurable instability, on which I was working at the time of my last meeting with Kurchatov, brought a painful and emotionally hot conflict with Leontovich. The first draft of the article describing the findings by Rudakov and me was passed on to Leontovich for approval. He returned it to us with some suggestions on how to improve the final presentation. With these in mind, we spent a couple more days revising the manuscript and went to see Leontovich again. When we brought it to him, he was not satisfied with the second version, and made a few more comments.

I don't know how it happened, but my temper, combined with the maximalism of youth, suddenly exploded. Refusing to introduce any further changes or even to talk to Leontovich, I went on strike. Two days later I met Artsimovich, who told me how much Leontovich was worried about this sudden tiff between us. Leontovich had confided to Artsimovich that it was especially painful for him. He had apparently considered me his heir in plasma theory.

A day later I heard the familiar sharp coughing in the corridor. The door opened and Leontovich proposed a cease-fire.

Many years later, when I was raising my own pupils in theoretical physics, I understood that what Leontovich had done was not a concession at all; it was a victory of wisdom and experience over youth's arrogance and stubbornness.

My candidate's diploma and the blessings of Kapitsa and Landau came during my last year in Moscow at the Kurchatov Institute. But at the time I had no idea that my time there was running out. The controlled fusion work at the institute had received international recognition and had become the object of great pride for the authorities. There was a rumor in the institute that the highest symbol of Khrushchev's benediction was to include the institute in the visit of President Dwight D. Eisenhower during the forthcoming summit in Moscow.

Kapitsa remembered one of his last meetings with Kurchatov, who as early as January 1960 was busy with preparations to meet the important guest.

"So, what are you going to do specifically?" asked Kapitsa.

"I have instructions from Khrushchev to impress the American president," responded Kurchatov.[3]

Of course, issues of scientific cooperation, even on Atoms for Peace, would never be the main items of the coming summit's agenda. Quite understandably, the leaders of the two superpowers had to discuss how to reduce the risk of nuclear war. Since it was in the very early stages of arms control and international security negotiations, this art hadn't yet reached the level of sophistication as with the SALT (Strategic Arms Limitations Talks) or START (Strategic Arms Reduction Talks) treaties much later. Both sides suggested several innovative ideas.

In brief, both sides were ready to talk about certain thresholds to limit the accumulation of deadly thermonuclear arms. One of the trickiest parts of any deal is the issue of how to control the agreed limits. With a distance of only seven years since the death of Stalin, confidence between the superpowers was far from established.

Eisenhower suggested an approach to verification called the Open Skies concept. It assumed that on agreed-on dates reconnaissance air-

3. As told by Sergei Kapitsa.

planes of both sides would fly along previously established routes over each other's territories.[4]

Khrushchev, in his turn, did not much like the idea of "open skies." It was clearly incompatible with the Soviet mentality of a closed society. So, the situation was rather unpredictable.

On May Day 1960, Khrushchev was catapulted from his bed at 5:00 A.M. by an emergency call. An American high-altitude reconnaissance U-2 plane had just entered Soviet airspace. He was bewildered and tried with great difficulty to hide his fury while nervously waving his hand at demonstrators in the May Day parade on Red Square. He waited impatiently for the final report from the Air Defense Command. The message on the downing of the U-2, and the capturing of pilot Gary Powers, was delivered to Khrushchev on top of the Lenin mausoleum.

Khrushchev felt humiliated by America's unilateral approach to Open Skies. The summit was canceled by the Soviet side. Who knows what the dominant factor had really been. Was it the lack of patience and wisdom on the part of Khrushchev, or was it based on a decision to postpone summitry until a new and friendlier president took the Oval Office?

The Soviet press was full of laudatory stories on victory in the air over Sverdlovsk, the place where Powers's aircraft was intercepted. Common people, however, were not particularly happy with the cancellation of the American guest's visit. The nation kept warm memories of a much more important victory jointly achieved over the mortally dangerous common enemy—the Nazis. The name of General Dwight Eisenhower, the hero of that war, had not been forgotten.

For us nuclear scientists, it was especially clear how important that missed summit could have been. Who, if not the atomic scientists, was most conscious of the dangers of nuclear war, based on the potential use of their own creations? Even if our concerns hadn't stopped us from making the atomic bomb, gradually we had begun more and more to assess the dangers associated with the discovery of radioactivity.

Sometime during 1960, we scientists discovered that Leontovich had somehow become persona non grata. He lost his privilege to travel abroad. The Iron Curtain did not open for him. Though Beria had gone,

4. Reconnaissance or verification from space, which would not require any agreement from the opposite side, was still many years away.

his ghost was still there and the promise he had made at the time of his appointment—to vigilantly watch Leontovich—had been fulfilled.

This interpretation aside, we took such turns of fate as something independent of our own will and sent to us from a different, nonphysical space. Bad luck or good luck, everything depended on chance.

Despite Leontovich's misfortune, for the most part we were happy in our internal exile, in the environment of exciting scientific discussions and discoveries, formulas and theories. I personally had no reason to complain, working in close collaboration with my two brethren, Yevgeny Velikhov and Aleksandr Vedenov. We formed a famous trio, developing what we thought was an interesting and far-reaching approach to predicting the behavior of plasma, a substance that is subject to incurable instability. According to our theory, the buildup of unstable motion in plasma, like the spontaneous excitation of different types of waves, signified transition of the plasma to the state of chaos.

I had a distinct feeling at that time that massive theoretical brainstorming could lead to a real breakthrough. As a result of such an offensive, plasma physics could eventually find a way to live with incurable instabilities in magnetically confined configurations of hot plasma. If chaos were unavoidable, then let it be controllable chaos, in which fundamental processes destroying the plasma confinement, such as heat conduction or losses to the wall, could be predictable.

From what science has now learned about chaotic dynamics, I can see how pioneering our particular work was in 1961.

14 / Farewell, Moscow

➢

In 1959 I began teaching at Moscow Energy Institute. This assignment gave me a chance to communicate with young audiences. On these occasions I had put my eye on a couple of bright sophomore students. I knew that they would join me someday as active scientific bayonets. Somehow I felt I was approaching a time of change in my life, which came with a call from Andrei Budker, whom I remembered from the time of my diploma work at the university.

On my first encounter with Budker, I hadn't paid much attention to his invitation to join his team. I was then too preoccupied with my plan to join Landau's group. The glory of Landau had so overshadowed everyone else that I had no intention of learning more about Budker and what he was doing in physics. Later, when I became a member of the Kurchatov Institute, I discovered what an unusual man and scientist Budker was. During my time at Kurchatov, he led a rather modest-size team that developed new ideas for particle accelerators. My friends in fusion told me that his innovative ideas in that field bordered on science fiction. Soon I discovered his impact on the controlled fusion research I was doing.

A couple of years after the pioneering work of Sakharov and Tamm, Budker suggested an alternative approach under the code name Corkatron. The elegance of the invention was in its use of stronger magnetic fields on both ends of a so-called magnetic bottle. The hot plasma in such a "bottle" would be protected by magnetic "corks." Kurchatov obviously liked Budker and his ideas. The Corkatron project was given a budget and eventually a large group of experimental physicists and engineers. In one of the laboratories in the institute they even tried to build an actual plasma machine to implement Budker's idea.

It happened that my very first discovery of plasma instability had been made precisely for this device, Budker's corkatron. After identifying this particular phenomenon, I was asked to deliver a talk at the Artsimovich seminar. It was understood that the presence of plasma instability in Budker's invention threatened the very viability of the project.

The huge room was overcrowded. Such an attendance, as I discovered somewhat later, could be explained by the psychological drive so characteristic of the scientific community. They were expecting a clash, and there was something bullish in the overall desire to witness it. Fortunately, in science it is usually a fight between different ideas and interpretations. Nevertheless, as Landau would say, people "sensed the smell of frying."

Budker, of course, was invited. Everyone expected an exciting duel between the two of us. I delivered the speech, and my arguments were rather straightforward.

When I had finished my talk and answered a number of questions, mostly of a technical nature, everyone expected a violent reaction from Budker. Instead, he stood up and generously congratulated me for "a very interesting contribution."

Obviously, many of the attendees were disappointed. However, Budker and I, from that time on, became good friends.

In the coming months and years I had a chance to learn much more about him. He came to Moscow a few years prior to World War II looking for admission into the Department of Physics at Moscow State University. A descendant of a poor Jewish family from a rural part of the Ukraine, he was full of ambition and self-confidence.

After the war Budker developed into both an excellent theorist and an engineering and design genius. "The relativistic engineer" is the characteristization that best fits him. He had a great flare for making quick estimates, as physicists call them, on the backs of envelopes or paper napkins. He was ever ready for the eternal intellectual adventure.

In his Kurchatov years, he made important contributions to the theory of nuclear fission reactors and then to plasma physics and controlled fusion. Finally, he built his own small empire within the institute in the field of high-energy accelerators. Some of the principles he developed in that area are still relevant in such contemporary projects as the superconducting supercollider.

Budker tried to invent more economical and elegant approaches to the basic experiments in high-energy physics. For the time being, Kurchatov gave him a chance to play all these games of inventing in a rather small place under the umbrella of the institute. It was a kind of incubator, where Budker first built miniature accelerators to test his ideas and wait for the moment when he would sail with his team into the open ocean of big science.

Such a moment came one day in the late fifties when Khrushchev's government adopted the project to establish a completely new center for science in the heart of the Siberian taiga. While the original idea had been suggested by the prominent hydrodynamicist and applied mathematician Mikhail Lavrentiev,[1] Khrushchev immediately supported it, since he felt irritated by the overkill of concentrating big science in Moscow. Blessed by Kurchatov, Budker joined the group of founding fathers of what was soon called Akademgorodok, which means "Academic City." Planned as a new campus, it was built twenty-five miles from the big industrial and cultural center of Western Siberia, the city of Novosibirsk.

Budker started recruiting future teammates for the Siberian adventure. He was very persuasive when talking to quite a few of us.

"Gentlemen, we need intellectual creative freedom, open space. You need a new life. After all, you could escape from these communal micro-apartments." I did not reject the offer bluntly. I tried to compare what he was promising with what would happen to me if I continued to stay at the Kurchatov Institute, where I had already established myself as a respected member of a team. I liked most of the people. The institute was also one of the world leaders in fusion research. In a new place, I would have to start over from the beginning, even if my material life would be much better. Maybe Budker was simply the devil, the tempter, I thought. Such were my vacillations.

Akademgorodok was clearly designed as a paradise for the hard sciences. But what could it offer the members of our families? our wives? The founding fathers had in fact envisaged even such details. With this in mind, I was almost in a position to persuade my wife to risk moving. Before the final decision was made, however, one unusual episode finally helped me make up my mind.

One day, after one of my first lectures at the Moscow Energy Institute, I was approached by a young man whose face looked somewhat familiar.

[1] No relation to Oleg Lavrentiev.

He said, "I am Volodia Zakharov. Do you remember me from Kazan? I am the brother of your friend Yuri."

Zakharov joined my seminar in theoretical physics. From time to time I gave him different problems to solve. He was especially bright in mathematical problems. Projecting into the future, I could see Zakharov as a member of my team.

Suddenly, in a state of complete desperation, he came rushing one evening to see me.

"Roald, I am in big trouble and don't know what to do. There is a very high risk that I might be drafted into the army."

"But it is impossible!" I exclaimed. "All students have a special deferment."

"Not anymore. I was kicked out of the institute."

I couldn't believe it. He was considered one of the brightest seniors. His expulsion, however, was not related to his academic performance. Apparently he had become a hero, and at the same time a victim, of a rather bizarre story.

One night, he was awakened in his room in the students' dormitory by his girlfriend who was in a state of great panic. She told him that after a party another student had tried to seduce her. In a state of extreme rage and agitation, Zakharov immediately jumped up, found the guy, and apparently beat him severely.

I agreed to find a way to help Zakharov. In a fraction of a second, I figured out that the only one who could save him was Budker. The next morning, I rushed to see Budker.

"Dear Andrei Mikhailovich, I have to tell you that one of my brightest young students has gotten into big trouble and only you can help him."

Budker invited Zakharov to tell him the story. After hearing the sordid tale, Budker gave his own assessment: "Volodia, I believe that you behaved like a real man. For a man who was—and maybe still is—in love, what you did was quite understandable. But if you were also an intelligent man, you would have found a different solution. An intelligent person would never resolve a dispute with brute force. Now, let's talk about your future."

That is how Zakharov was sent as a technician to Akademgorodok. While I was still in Moscow, I had to accept the responsibility of providing assignments to keep this "technician" busy performing all the tasks typical of a theoretical physicist. That's how I was finally—simply by force of circumstance—linked to that distant campus in Siberia.

15 / Novosibirsk

➢

Going to Novosibirsk was like taking part in a huge experiment—reminiscent of the nostalgic feelings for the heroic epoch of Arctic exploration. The difference was that we did not land on an iceberg. We came to a modern town that was sharply different from the mediocre living conditions and environment typical of such Siberian settlements. The governmental decree to launch Akademgorodok injected 220 million rubles into the project—a substantial sum of money for the late fifties. With Khrushchev's personal signature on the decree, he established himself as the grandfather of Akademgorodok.

For some reason, I missed Khrushchev's first visit to Siberia for the first on-site inspection of the new city. His tour gave an additional boost to the construction work there. However, unfortunately for the future dwellers of this campus, Khrushchev also left the memory of his architectural ambitions. He didn't like the blueprints for the largest building in town, the hotel for guest scientists. He thought the building would be higher than the trees of the surrounding taiga forests. To improve the building, from his point of view, was very simple. He took a pencil and cut the building in half. From that time on, the organizers of big international conferences or seminars faced a shortage of hotel rooms, and they blamed the "great architect" of the post-Stalin era.

All of us in Akademgorodok were waiting for Khrushchev to come a second time. Fourteen institutes and a newly opened university on the campus were in a state of alert, making last minute preparations. Mikhail Lavrentiev, president of Akademgorodok, drove his jeep from one institute to another to attend the rehearsals. Khrushchev, he said, had to be impressed with the science we were offering him.

In Budker's Institute of Nuclear Physics, the principal show, of course, had to be the introduction of particle accelerators to our most important guest. Budker knew, however, that it would be difficult to impress a layperson with the electrons and positrons revolving in the accelerator's vacuum vessel. They were practically unseen. Thank God it was the time of the early boom in lasers.

The standard ruby laser provided a spectacular demonstration of how a concentrated light beam, in a pulse as short as a billionth part of a second, makes a hole in a coin used as a target. The commission responsible for the rehearsal was satisfied. But as the only theorist among them, I thought that something was missing.

"Gentlemen, you are wrong to use a soft target, I mean soft currency. If we really are to impress Khrushchev, we have to find a piece of hard currency."

I was very proud when this very modest recommendation was approved. Someone produced a quarter dollar from his recent foreign trip.

After the successful rehearsal, everyone waited for the day when Khrushchev would turn up. We all thought that there would be nothing to interfere with his trip. But Khrushchev did not come. At the trip's scheduled time, the Cuban Missile Crisis erupted. With his inability to come, Akademgorodok lost forever the chance to impress Khrushchev. Khrushchev did not have much time left before he was deposed as Soviet leader.

The life of the little academic community in our town was exciting and comfortable in a straightforward, earthy way. Even in the rainy Siberian autumn, we were safe without high rubber boots, walking on asphalt-covered pathways between birch trees. We felt like czars in our enormously big (as we saw them) apartments. The food supply also came from the most privileged warehouses of the Ministry of Medium Machine Building, the ministry that ran all the nuclear installations in the country. Akademgorodok, for quite a few years, enjoyed all the benefits of the classified atomic enterprises without being surrounded by barbed wire.

While people in medium-size Novosibirsk stood for hours in line for butter, meat, and sometimes even bread,[1] we, the residents of Akademgorodok, were getting regular home deliveries of foodstuffs. Such privileges contributed to the growing irritation of the population around Akademgorodok. The local government and the regional party bosses reflected the hostile feelings against the scientific community. But they

1. This happened in the autumn of 1962, the first year of Khrushchev's grain crisis. From this year onward, the Soviet Union became the world's biggest importer of grain.

used these complaints to taunt us for our freedom of spirit, and for the different moral and intellectual climate we were cultivating in Akademgorodok.

A local party boss, the regional party secretary of Novosibirsk, Fyodor Goryachev,[2] was able with great reluctance to tolerate Lavrentiev's independent status. Goryachev tried to restrain himself when the protective shield, extended by Khrushchev, was evident. But the instinct of this professional politician immediately sensed that Khrushchev was no longer visiting the campus to bless this special "scientific city." Khrushchev was busy with escalating internal difficulties, as well as a series of international crises. He had barely come out of the Cuban Missile Crisis, when the deterioration of Sino-Soviet relations delivered another strong blow to his regime.

Goryachev took this opportunity to move against Lavrentiev. We felt it in our everyday life—petty commissars often came to inspect the ideological life of the institute. More and more the local press actively attacked us. But Lavrentiev was vigilant; and he, too, was a professional of high politics. Unable to set up an appointment in Moscow with Khrushchev for quite a long time, Lavrentiev took the unusual risk of intercepting Khrushchev during an important state visit to the People's Republic of China. Landing in China almost immediately after Khrushchev's delegation, he appeared before the boss at a chosen precalculated moment and seized Khrushchev's attention by complete surprise.

"What are you doing here?"

"Dear Nikita Sergeyevich, I have a very important message for you. Please, may I fly back to Moscow with you? I need this chance, desperately, to talk to you."

The request was granted and, as a result, the siege of Akademgorodok was called off. Khrushchev would not give up his beloved child, which in his table of ranks was second only perhaps to the great dream of the development of the virgin lands in Siberia and northern Kazakhstan.

The scientific community was no stranger to Siberia, even before our campus was established. After all, how many of the best minds of the country had been sent to Siberian gulag camps during Stalin's purges?

2. Goryachev had the reputation of a hard-liner even by Khrushchev-era standards. For several years he was the boss and mentor of Yegor Ligachev.

The scientists of that generation were kept in an environment nobody would envy. Very few survived until the golden valley in Akademgorodok could extend its hospitality. They joined us as strangers from a different, almost nonphysical, space, not simply from the continent that Aleksandr Solzhenitsyn called "Gulag."

One of the rare lucky survivors of the camps, Yuri Rumer, became a very dear friend of mine. As a talented young linguist already capable of speaking a multitude of Oriental languages, Rumer fell in love with theoretical physics. Eventually he won a fellowship for postgraduate study in the mecca of European physics, Goettingen University in Germany. Max Born, a Nobel Prize winner and one of the founders of quantum mechanics, took Rumer under his supervision. Many of the best brains in physics were pupils of Born.

A few years later, Rumer came back to Russia as one of the hopes of young Soviet science, bringing with him strong ties and new friendships with those few dozen people who had defined the skeleton of twentieth-century quantum physics. But he could not have known that he would never be able to see them again.

With Landau, Rumer established not only a close personal friendship, but also a plan for long, fruitful cooperation. Landau launched his monumental project to write a series of textbooks on theoretical physics. Rumer volunteered to translate some of the most complicated ideas of contemporary physics, like Einstein's theory of relativity, for a much wider Russian audience of laypersons and pedestrians.

The comrades' plans were destroyed by the purges in 1938. They were both arrested—first Landau, and then a few days later, Rumer. The authorities tried to accuse them of one and the same crime. Rumer was treated as a coconspirator and an accomplice to Landau, the principle suspect, who had "tried to misinterpret the principles of modern physics to inflict damage on the socialist economy."

Both were expecting a merciless sentence. The miraculous rescue of Landau by Kapitsa through direct appeal to Stalin brought hope to Rumer. Poor fellow. Rumer did not know then that the irrational logic, used by Beria and his repressive apparatus, had nothing in common with the basic wisdom of physics, the principle of causality (the cause must precede the effect). While the principal criminal was "liberated" and declared not guilty, the person labeled as his accomplice was sent for many years to Siberia.

Instead of being immediately and irreversibly sucked into the black hole of the gulag, Rumer was brought to one of the *sharagas*[3]—the closed enterprises run by Beria's department. This particular *sharaga* collected at that time a most unusual group of people who were predestined many years later to open the Soviet space program with the launch of *Sputnik I*. The *sharaga* housed such future space luminaries as Sergei Korolev and Valentin Glushko. But now, in that closed environment, they had to work as forced intellectual serfs on the design of new warplanes.

A few years later, when aircraft and rocket engineers were critical for Stalin's plan of military buildup, they were released and brought back to Moscow. But Rumer got a kick in the opposite direction. Unnecessary for that particular technical push, he was exiled to deep rural Siberia, to a tiny town lost on the banks of the Yenisei River.

As a political outlaw, he had to register his presence every week at the local police department. His only support at that time came from his faithful and loving wife, Olga. A free citizen, she shared the ordeal of exile with her husband. A simple Russian woman, she kept a lifeline, a bridge, connecting Rumer with his beloved science. Her care and material support gave him, despite his conditions, the chance to practice theoretical physics.

It would probably be difficult to imagine a more hostile environment in which to work on the sacrosanct problem of theoretical physics, the one that was an obsession and unresolvable challenge even for Albert Einstein.[4] In the somber light of his kerosene lamp, Rumer finally produced his own version of the unified field theory. He thought that getting beyond the four-dimensional space of the theory of relativity[5] would be the key to success. The fifth dimension is what physicists call "action," which gives new life to a familiar physical quantity.

Rumer's calculations were straightforward and elegant. He truly thought it was a breakthrough that deserved to be communicated to the world of free scientists. But the only way to break through the wall separating him from that world was to write a letter to Comrade Stalin.

3. A special form of "intellectual" gulag camp or prison, where the inmates worked on defense projects.

4. For twenty years of his life at the Institute of Advanced Study in Princeton, Einstein tried to construct this superscience, which, like monotheism in religion, encompasses and brings together fragmentary theories of different kinds of forces in nature—nuclear, electromagnetic, and gravitational.

5. In which time is added to the three conventional spatial coordinates.

Rumer recounted to me that the most difficult part of the whole story was to figure out how to address Stalin. Kapitsa, even under house arrest, still could continue in his own way: "Comrade Stalin." But how could Rumer, a political outlaw, almost an "enemy of the people," call Stalin a comrade?

A thick formal letter enumerating the most important official titles of the addressee was sent to the Kremlin. It contained the theory of five-dimensional physical space proposed by this scientist in disgrace.

Clearly, the substance of the paper was the object of examination by experts, Rumer's former colleagues. For Stalin's consideration, Rumer had brought only a general, rather philosophical argument—he quoted Lenin. In one of his prerevolutionary exiles, Lenin had spared time for a philosophical analysis of early twentieth-century physics from the point of view of Marxism. While Lenin in general had praised the revolution in science, and especially in physics, he didn't much like the use of mathematical abstractions. He even remarked sarcastically, "Four-dimensional space has not helped any obstetrician."

In his letter to "the only living classic of Marxism," Rumer exclaimed: "I volunteer to be such an obstetrician!"

The officials in the Central Committee of the Communist Party of the Soviet Union (CPSU) sent the manuscripts for peer review. Several leading theoretical physicists were asked to express their opinion on the importance of the new theory. Such an assessment was a dramatic litmus test.

One of the reviewers was Vladimir Fok, himself rescued by Kapitsa from the Lubianka in 1937. He read the paper, acknowledged the elegance of Rumer's mathematical construction, but said that it was rather transparent for him that the theory was not any better than most of the suggestions already published in physical literature. Nevertheless, in Fok's evaluation the paper passed the examination and the test for decency, integrity, and generosity. However, not all of the reviewers made the same evaluation.

No one knows how the final decision was made, or whether they actually carried a ballot between the reviewers, but Rumer was saved. The authorities permitted the publication of his article in a physical journal and gave him the right to resettle in Novosibirsk. That was just a few years before the huge crowd of Moscow scientists arrived on the nearby taiga.

Rumer, in the meantime, was promoted to the post of director of one of the local institutes and was given the chance to advance his theory. For more than thirty years after that time, he had the happy creative life of a physicist. Though he never again mentioned the theory he had sent to Stalin, the "action" remained dear to him to the end of his days, perhaps as a memory of his extraordinary intellectual escape into the fifth dimension: his flight from the cruel realities of the totalitarian regime.

The story of another survivor of the gulag, Vera Lotar-Shevchenko, is no less breathtaking, though it developed in a different dimension. Only in the end did the labyrinth of life bring her to Akademgorodok.

Lotar was a talented French girl, a graduate of a conservatory, who had promising prospects in concert life. But she fell in love with a handsome Russian émigré. They decided to unite their lives. However, the young husband was obsessed with the idea of repatriating to Soviet Russia. Lotar supported this intention. After all, Russian musical culture was world famous.

It would have been impossible to find a more inappropriate moment to go back to Russia than on the eve of the Great Purge. These young people were barely given the chance to get accustomed to their new home when Shevchenko was arrested by the KGB.[6] Lotar was lost, with no idea of what to do in this enigmatic Slavic country. She knocked on all the doors and visited all the offices in an attempt to understand what might have happened to her poor husband.

Finally, the authorities decided to get rid of the importunate stranger. She was sent to one of the gulag camps in a distant part of Siberia, infinitely far from her own *patrie*. Her understanding of the situation came soon. Along with thousands of her camp mates, she was forced to shovel the frozen soil as a slave.

The regime sought to kill hope in the souls of its victims, but the soul of Vera Lotar-Shevchenko belonged to music. Overcoming the unbearable tiredness from exhausting forced labor, she transcended to the mentally created world of sound—hour after hour, month after month, for years. Every day after shoveling, she put her swollen, hardened fingers into action in her imaginary world. It could well have been a kind of self-inflicted hypnosis, in which she moved her hands and fingers as she

6. It was a rather typical fate of those émigrés who came back to Stalin's USSR.

played Chopin or Beethoven. All of her favorite musical repertoire for piano was recreated.

She was faithful to music, and music helped her to keep her faith—the name *Vera* means precisely that—and the confidence that one day her hands would again touch a real keyboard.

In Akademgorodok she was musician in residence, adored by all of us. People couldn't stop crying after her concerts. We were moved by her sound and her spirit.

With the quickly growing scope of international contacts, foreign trips, and the intensive influx of international guests attracted by the fame of our Siberian establishment, there was an obvious need to have a chance to practice English in a natural way without formalized lectures or exercises. We decided to start something like a club where people could go to talk to each other in English on different subjects. We planned to invite the most interesting guests to deliver nonscientific talks.

The presidents were chosen on a rotational basis. Fortunately, my turn came after the club had already been established. Thus, I had been able to avoid the embarrassing situation of not being able to communicate with my constituency in the language declared as "official" by the charter of the club. My active vocabulary of English words indeed was very poor in the beginning.

On my very first trip to the United States from Akademgorodok, I remember I had to enter a barbershop. The polite Bostonian barber seated me in the chair and asked me a question that I was unable at that time to fully understand. But somehow I had the feeling that he wanted to know what style of haircut I would prefer. A desperate search of my English memory files in a few seconds provided an answer: "Medium."

Once in late autumn, the English club members started lobbying for a Christmas celebration. We wanted to recreate the atmosphere of apple pie and to sing appropriate Christmas songs. By the end of the preparations we had made huge progress. We even found a room with a fireplace in the faculty club of Akademgorodok. But then, suddenly, we sensed the smell of danger. The district committee of the Communist party, which controlled the ideological life of the town, vetoed our undertaking. Embarrassed and surprised, I called the secretary of the party committee to find out what was the matter with the idea.

"You must be completely crazy! Your idea is acquiring undesirable political coloration!" I was told.

I replied indignantly, "Not at all! We are planning to reconstruct a simple apple pie Christmas party in a home-style atmosphere."

The secretary then launched into what he probably thought was a provocative suggestion: "Okay, if it is a simple, innocent home-style party, why can't you move your celebration from this suspicious date, the twenty-fifth of December, to any other date?"

I had only a fraction of a second to respond. "Esteemed Comrade Secretary of DistriktKom, I tried. But it is virtually impossible. Look, the twenty-first of December is the birthday of Stalin. The twenty-sixth of December is the birthday of Mao. The Capricorns are terrible! We have no other choice."

The party commissar retreated with a laugh. "You bastards are too bright. Okay, you can celebrate the birthday of Jesus Christ, but I beg you, please, no publicity."

16 / The Giant Leap Forward

➢

While cooperating on the creation of an oasis of science in Siberia, government and scientists had had rather different agendas. The latter, including my fellow physicists and me, had hoped very much that by coming to virgin land we could start everything from scratch, according to international scientific standards, instead of waiting for God-only-knows how long in Moscow's old established institutes. We wanted to catch up with the West.

The encounter with our foreign colleagues at international conferences, however, was an eye-opener for us on a new style of doing science, free from the boring and even intimidating constraints established by the administration and the bureaucracy; free to travel when we ourselves felt it was important; and free from fortified fences around the perimeters of mailboxes. We even hoped that the irritating ideological indoctrination that came from the party hierarchy would somehow miraculously evaporate.

The Khrushchev government, in its turn, considered the massive transplant of scientific excellence to Akademgorodok a great way to boost the technological and cultural development of Siberia. The establishment of a new academy would produce a scientific community more dynamic and responsive to requests from authorities. The new academy, they thought, would consist of specially selected and purified scientists.

From the very beginning, it was obvious that the inhabitants of the new town in Siberia would be from the young generation of scientists. The laws of natural selection dictated that they could move from Moscow and Leningrad with greater ease than those who had already built their careers and a comfortable life in other established institutions. If the official master plan indeed consisted of moving forward the generation of scientists brought up by Soviet power, nominally no one could have had better credentials than Andrei Budker.

Budker was born into a poor family in the year of the "Great October" (1917). Soviet power opened the door for him to Moscow State University. He graduated at the peak of Stalin's great Five-Year Plan to transform the Soviet economy. He fought in the war against the Nazis and met Victory Day as an artillery captain decorated with many medals. He received a few additional awards as a prominent member of Kurchatov's team. When his Socialist homeland sought to become a nuclear power, there was, of course, one "small dark spot" against him: the sound and spelling of his name.[1] The government, however, was generous enough to make an exception in recognition of such outstanding service.

Within the new academy, Budker's style was to nourish a team of like-minded scientists, and for a rather long time I was one of them. He shared with us all of his ideas—his hopes and his doubts. In return, he requested complete dedication, even self-sacrifice.

Our alma mater in Novosibirsk—Akademgorodok, the Institute of Nuclear Physics—was built on the principle of collective brainstorming. Budker had ordered the institute's workshop to design and produce a rather fancy round table that could comfortably accommodate a couple dozen young scientists and their director, their leader. The very shape of the table symbolized the spirit of democracy in scientific discussion. Budker himself liked to compare it with the Round Table at which King Arthur had conferred with his brave knights.

If Kapitsa was an introvert, concealing his scientific ideas and engineering innovations until they could be developed and given final elaboration, Budker in contrast was the true personification of an extrovert—he liked to think aloud. For that reason, we were lucky to be not only witnesses, but also a part of his inner "kitchen" while he was cooking a new and quite often bizarre feast.

Budker needed criticism and suggestions in order to elaborate new ideas jointly. But sometimes he behaved like a grown-up kid, offended by the remarks of skeptics. "It is so difficult to find something interesting and new, when instead of helping me develop fledgling ideas in their infancy, you fellows immediately look for a way to attack."

At the same time, Budker was very proud of his boys. The first years of the institute were tremendously successful. By 1963, we had already built our first particle collider, an early prototype of the future great machines of high-energy physics. In that collider, VEPP-2, a beam of

1. Being of Jewish origin could have worked against Budker when it came to sensitive scientific responsibility.

electrons revolving in a vacuum vessel under the force of a magnetic field was made to collide with a counterstreaming beam of positrons, the alter ego of electrons in the world of antimatter.

The creative atmosphere in the institute was a direct result of the continuous effort and attention Budker paid to the issue of hiring people and working with them every day. I do not remember a single instance when somebody was appointed to one of the science teams without first being interviewed by Budker himself. This was important for the stability of the institute. Those who were finally selected had the huge luck not only to work in a very exciting place, but also to be taken care of by the director, Budker himself. He helped in every respect, from the allocation of living apartments, to finding a job for someone's wife, to getting the little kids into kindergarten.

His personnel policy and his tireless efforts to communicate daily with team members prepared us well. We were ready for major undertakings and anxious to compete with the best science teams in our country and abroad. Budker liked to say with pride, "We have a superiority of bright heads." No one could counter his statement. All of us accepted the rules of the game in which Budker was the enlightened czar. The rest of us, in moments of greatest dissent, were "loyal opposition to the king."

In order to take care of his institute, Budker had to deal, unfortunately, not only with the faithful knights sitting at the round table, but also with the external world, controlled by a whole hierarchy of bureaucracies not always so friendly to science.

To overcome the potentially hostile feelings of the local establishment in old Novosibirsk city—feelings based on envy and jealousy toward the newcomers from privileged Moscow—Budker appointed one of the local apparatchiks as his first deputy. It was a marriage of convenience, and Budker's choice was made very carefully and thoughtfully. The man's name was Aleksandr Nezhevenko.

Formerly a member of the regional party bureau in Novosibirsk and the leader of some of the biggest industrial enterprises in the area, Nezhevenko soon became an ardent supporter and promoter of our science. Years later he confessed that he had never been so happy before in his life. For Budker and the institute, it was essential that Nezhevenko, while falling in love with science, not lose the fighter qualities he had acquired during his rise in the party hierarchy.

Budker paid special attention to the need for continuous education and reeducation of VIPs, on whom the fate of the institute depended. He never regretted spending hours of his time talking to important guests to explain the ideas of colliding beam experiments, which were destined to become a principal instrument, a kind of looking glass for the subatomic world.

Very soon, the institute had the chance of presenting to our guests not only lectures, but what was more important—live demonstrations of the subatomic world in action. The very specific light emitted by the fast-moving charged particles in a magnetic field—so-called synchrotron radiation—became the principal attraction of our scientific Disneyland for distinguished guests.

The supervisors from the Central Committee in Moscow or from the Ministry of Medium Machine Building, whose support was essential for very important large-scale nuclear physics projects, often came for much longer visits. Budker never hesitated to tempt them with the chance of becoming part-time researchers, at least nominally, or of producing the theses of dissertations: "Why not have our men in higher echelons of power? The academic regalia would not spoil our friends. At the same time, they can use them for further promotion."

All of this campaign—of lobbying, entertaining, and sometimes schmoozing with the big bosses—was only a prelude. It served as the precursor to a major undertaking that Budker conceived: to get "approval" and to codify the status of "most favored institute."

He started a long series of sorties to Moscow to make believers out of government officials—that is, to turn them into supporters of his science. It was not an easy task. Often he came back to Novosibirsk rebuffed, once again rejected, but not humiliated, and certainly not defeated.

He would tell us, "I have a philosophy. If I'm rejected today, I will try again tomorrow."

These frequent trips to the Central Committee, to the State Planning Committee (Gosplan) and so on were an eye-opener for Budker. He discovered that in the most totalitarian, autocratic, and centralized system, the real power was concentrated in the hands of an invisible army of low-key clerks, who occupied less prestigious offices than those of members of the Politburo and ministries. These clerks played the key role in drafting the most important documents and pushing them through the bureaucratic labyrinth before the papers even got to the desks of the biggest bosses for final blessings. These fellows knew much better how and when to approach the boss, and what to tell him to change his mind.

This is how the institute was given the special privilege of selling innovative hardware, like electron accelerators for industrial use, not according to very strict governmental regulations, but by watching the market. Whenever demand exceeded supply, supplies had to be reestablished in negotiations with potential customers, the clients. The institute, with these breakthroughs, was launched into a higher, more privileged orbit in the interest of doing better science.

Being part of all the discussions and witnessing Budker's efforts, I was quite supportive of the final results. However, I felt very sorry for him with respect to the substantial overtime he had to put in to run the institute.

"There must be something deeply wrong," I thought to myself, "with a system that so inefficiently uses its scientific brains." I could hardly know at that time that I myself would not escape the same fate.

One particular episode in Budker's entrepreneurial activity brought the first serious disagreement, if not an open conflict, between us. Budker was in Moscow, and we were expecting his return with another important visitor. "Have you ever heard the name Vasily Mishin?" asked Budker. "He is the direct heir to Sergei Korolev."

All of us at the round table considered the visit of the rocket czar with great curiosity. Even more, we were intrigued by the reason for his interest in our institute, whose topics were so remote from those of the space program.

One day in the autumn of 1969 the corporate jet of the former Korolev company delivered Mishin and Budker to Novosibirsk after four hours of nonstop flight. The legendary Soviet spaceman disappointed most of us. Instead of the inspiration we expected to radiate from him, he smelled of spirits in overabundance.

The brainstorming that followed the arrival of Mishin and Budker was about, to use contemporary terminology, a strategic defense initiative. Our guests were extremely interested in whether or not the institute could design transportable particle beam accelerators that could be carried on their rockets in space. If sufficiently intense, such a beam could destroy enemy missiles. Even if it were capable of making a small hole in the wall of the target's fuel tank, it would be enough.

The meeting with Mishin, and the first direct encounter with issues of such high strategic military importance, had a strong impact on me. And it had big repercussions in the minds of the round table knights. Budker was interested in getting a huge contract to develop this idea. It was so much ahead of its time in 1968 that even fifteen years later, when Reagan suggested the Strategic Defense Initiative (SDI) in his famous "Star Wars" speech, the idea of using particle beams was qualified as exotic technology.

Inexperienced as I was in strategic analysis, however, I challenged Budker's approach on moral grounds: "How could we accept the contract when we haven't even tried to design the scenario of how such things work and what the final outcome would be? Is it fair to spend money before we have had a chance to even think seriously on the issue?"

Budker replied, "It is moral to think of the protection of the sky above us, above our country."

I tried to insist: "Okay, you want to protect it? Let's think about it first, and then when we have made up our minds about whether we can really do something about it, we can talk about the financial side."

"No," he said. "It would be foolish to spend mental energy thinking without being paid for it. After all, all these expenditures are but a drop in the sea of resources wasted for the military budget anyway."

The further development of this story, in my view, was quite fortunate. Assessment of the project, before it was to be implemented, was given to a number of peer reviewers. One of them was Artsimovich, who told me a year later that his role had been instrumental in killing the project. I didn't know then that it was at that very moment that the strategic thinkers and arms controllers of two superpowers had begun slowly moving toward a future antiballistic missile treaty that would consider such inventions potentially dangerous and destabilizing to strategic nuclear parity.

Unable to overcome Artsimovich's firm negative attitude, Mishin gave a rather modest contract to Budker's institute. I doubt it helped to advance the early SDI concept, but maybe it served the interests of high-energy physics.

Everyone who worked with Budker knew him as a great scientist and generous caretaker. This is how I will always remember him. I will never forget one of our long debates at the round table, when he was painting a rational scenario for further development of Akademgorodok. To do so would require the building of many more stores, boutiques, and kindergartens. Suddenly Budker added, "But we shouldn't forget about a cemetery."

The company of young men immediately burst into laughter.

"Don't laugh," he said. "But be prepared not only to live here, but to die here, too."

In 1976, I went back to Akademgorodok from Moscow as an official envoy of the academy, and as a friend of Budker's, to deliver the official eulogy at his funeral. I bade him a last farewell at that very cemetery. Budker was laid to rest—free at last from the decades of struggle to keep science alive under increasingly hostile conditions.

I felt proud that in those times with Budker, I was on his side of the barricade.

17 / No Bananas Today: The "Banana" Equation

➢

During the first year of my Novosibirsk period, I started to gradually build a similar atmosphere—to the one that Budker had created—with my own team. Zakharov, the heroic defender of female honor, was with me, as well as his classmate from the Moscow Energy Institute, Alec Galeev. Both had moved to Novosibirsk. By the end of 1961, I had ten people on my team. It was an incomparable joy to share ideas with others—with those who were like-minded.

Literature is an endeavor that is personal and intimate to the highest degree: it is by definition the way an author expresses himself. Science, designed to reflect the objective reality of nature and the surrounding world, has a collective cooperative character that is required to safeguard objectivity. Science can only be done by working in a team.

Controlled fusion, based on magnetic confinement of hot plasma, was precisely such a monumental field requiring the highest degree of collective undertaking. For a number of years from the early fifties until the early sixties, plasma physicists everywhere were deeply influenced by American theorist David Bohm's prediction of ultimate plasma behavior in a magnetic field. According to Bohm, a magnetic field could never do a proper job, and the plasma would escape and diffuse through the magnetic field fences too rapidly to be able to pick up sufficient temperature in the process of heating.

While many early experiments supported Bohm's prediction, theorists tried to understand what kind of physics might be behind this mysterious, anomalous diffusion, the formula for which was given by Bohm without any commentary or proof.

One of the early successes of my Novosibirsk team was the development of a quite plausible theory explaining this mysterious Bohm diffusion. Drift wave instability, as we called it, was discovered by my closest Moscow collaborator, Leonid Rudakov, and me under Mikhail Leontovich's supervision.

Time has passed since the early years of Novosibirsk, bringing a stream of elaborate and much more sophisticated models. But I am very proud that the framework we devised—to explain Böhm diffusion—is still there. It has survived the most difficult test in science: the passage of time.

Understanding the basic physics behind the phenomenon was the first step, and the second step was trying to play with the parameters of plasma and the magnetic field to see whether it was possible to cure the "disease," and to decrease the unacceptably high losses of plasma into the walls.

I was doing this type of thinking aloud with the then-young theorist from San Diego, California—Marshall Rosenbluth. Our friendship had started with the Geneva conference in 1958. But with every brief encounter at follow-up conferences, we discovered more and more our like-mindedness and synergism. One such brainstorming ended with the recognition that even slight but appropriate twisting of the magnetic field can eliminate or strongly suppress drift wave instability.

For Galeev, the youngest member of my Novosibirsk team—he was at that time only twenty-one—the elaboration of these arguments proved to be a serious test when, later in the autumn of 1961, I sent Galeev to Kurchatov Institute to report our findings at the Leontovich seminar.

Galeev came back in a few days with a letter from Leontovich: "Roald, you behaved like Balda sending your younger brother to the competition."[1] The letter from Leontovich went on: "The final score is five versus three in favor of the Akademgorodok team." This is how we entered an exciting, friendly, and mutually fruitful competition, not only between Akademgorodok and the Kurchatov Institute team, but in the worldwide scene.

At the beginning of the quarter, every department or lab had to make its so-called socialist promises or commitments, indicating how many more final products would be delivered on top of the regular state plan. Since

1. Balda is from a poem by Aleksandr Pushkin, "The Tale of a Priest and His Servant, Balda." The landlord originally perceived Balda as a kind of blockhead, easily cheated. However, Balda proved to be very smart at the crucial competition for fast running: he produced a rabbit out of a bag and threw him into the race as his younger brother.

the output of scientific activity could not be measured quantitatively, a method more appropriate for sausages, we always had difficulty with the authorities in formulating our promises.

The bright ones from the theory department found a way to ridicule the bureaucrats. "We have decided to commit ourselves to producing one scientific discovery of world-class standard, two discoveries of All-Union standard, and three discoveries of Siberian significance." Nothing could stop the physicists from joking.

April 1962 brought to us a veritable festival of science: Tamm and Zeldovich paid a special visit to Akademgorodok. Their formal excuse was to assist, as referees, the presentation of my doctoral dissertation. I had to agree with Budker, who considered such an academic milestone a boring formality.

In the process of preparing the final draft of my dissertation, I strictly followed Leontovich's advice: "Don't try to write the thesis. It would be immoral. It would steal invaluable, irreparable time from your creative scientific activity."

"So what should I do?"

"Very simple. Use the scissors-and-glue method. Bring together all the pieces from your published scientific articles."

Even though it was a formality, I was very happy that it provided the opportunity for the great men of physics to come to Novosibirsk. For a few days, they were with us, sharing our Bohemian creative lifestyle in Akademgorodok, including our bizarre late-night phone conferences.

Tamm concluded: "Don't worry. This is precisely what you have to do by the very definition of physics: physics is what physicists do when the working day is over." Zeldovich immediately picked up on the theme: "The only victims are the wives: ours are essentially the wives of sailors."

In 1965, Akademgorodok hosted the first representative international conference in theoretical and mathematical physics. Our colleagues were extremely intrigued by how quick Siberian science had picked up momentum.

A close colleague of mine and friend for many years, Harold Grad from New York University, expressed appreciation for what we were doing in Novosibirsk. Ready to make his own commitment, he said, "I will have to learn the Russian language." It was a time of rapid deterioration in Sino-Soviet relations. The era of cultural revolution was nearly upon us. I reacted in the spirit of the time: "You are an optimist if you learn the Russian language. The pessimists are learning Chinese."

My young men were maturing and eventually identifying their own preferred fields of competence in physics. Galeev was becoming a great expert in plasma physics. Zakharov, actively looking around for newer and newer fields of application for his talent in nonlinear sciences, came up with a theory of chaos in waves on the surface of the ocean. Alexei Fridman, their young roommate, found himself in the most distant of orbits around the spiral structure of the galactic disk. One of my pupils, Georgy Zaslavsky, came from the University of Odessa. Using our plasma background, we launched a study with him that may rightly be called the physics of chaos.

I myself devoted a great deal of time to talking to experimentalists. The Novosibirsk team, inspired by my theories of plasma shock waves, experimentally confirmed the existence of these waves and made an interesting observation of their generic relationship to solitons.

My friendship with Rosenbluth was strengthened in 1966, when we jointly chaired a think tank on plasma and controlled fusion in the newly created international center for theoretical physics in Trieste, Italy.

In the meantime, a most interesting experimental program was developing at my old alma mater, the Bureau of Electronic Equipment at the Kurchatov Institute. The bureau itself was renamed the Division of Plasma Research, and my old colleagues had moved into a substantially bigger building. The plasma devices grew to the size of a medium-scale power station. As a result, magnetic confinement gradually improved, and soon Artsimovich declared that there would be nothing resembling irreparable Bohm diffusion in his so-called tokamak device. The temperature of the plasma had already reached about 10M K (10 million degrees Kelvin).

The tokamak was a doughnut-shaped device, a much more advanced version of the original one suggested by Sakharov and Tamm. However, there was no theory to support Artsimovich's claim that a temperature of 10M K had been reached. The international plasma community was at first rather reserved in expressing its support for the announced high temperature. To counter this skepticism, Artsimovich made an unprecedented proposal: anyone could come to the Kurchatov Institute to conduct an "on-site inspection," if necessary, and his team would be ready to provide every bit of assistance should guests decide to carry on their own verification experiments.

A few weeks after Artsimovich's open invitation, a British team from Culham Laboratory flew several hundred kilos of hardware to Moscow. At that time, the most indisputable experimental technique to measure the temperature of plasma electrons was to scatter on them a beam of laser light. The hotter the plasma, the faster its particles moved. The frequency of the scattered laser light would then be proportionally shifted to a greater extent, like the whistle of a fast-moving train.

The final result of the experiment with a British-made laser beam proved the scientific claim of the Soviet tokamak team to be correct, and their presence at the Kurchatov Institute was a tribute to Artsimovich's new thinking and scientific glasnost and openness.

On the theoretical front, a breakthrough in understanding the tokamak plasma came with a paper I wrote with Galeev. We had decided for a while to get rid of the assumption that plasma was unstable and consider the best-case scenario of what would happen if plasma were stable. We came to the conclusion that the leakage of thermal energy from this type of plasma would be primarily associated with the existence of a special kind of plasma particles. Their trajectories in a magnetic field, corresponding to those of the tokamak device, would resemble the shape of a banana. Galeev and I even introduced the notion of a "banana" equation.

After some tricky mathematical operations, we attained a result, now known as banana diffusion. We brought it to Artsimovich. I remember very well the seminar in early 1968 at his lab. Almost everyone in his group remained skeptical, except Artsimovich himself. But most importantly, his sense of practicality and physical intuition overwhelmed his own tendency to be critically minded.

Within a few days, Artsimovich had done his own homework and successfully brought together the experimental data and our theory. Soon the banana theory was generally accepted, stimulating much more work in that direction by different groups of theorists. At least for a while, it was reminiscent of a renaissance of best-case scenarios that had always been the dream of classic thermonuclear fusion. This is why our theory was considered neoclassical. The name neoclassical theory is still widely used.

One of my old friends, Harold Furth, soon to become director of Princeton Plasma Lab, kept for a long time a huge banana-shaped flying balloon in his office. And as a sign of appreciation from the opposite side of the ocean, for a long time we in distant Siberia kept an American recording that had been given to us: "Yes! We Have No Bananas!"

While physicists were continuing their jokes, the world was gradually becoming a more and more dangerous place to live. I remember the shocking news of the invasion by Warsaw Treaty tanks into Czechoslovakia in August 1968. It was officially presented to us by Leonid Brezhnev as a fulfillment of socialist international duty. I had learned of this news on a plane flying from Novosibirsk to Moscow. I was busy collecting my documents for a foreign trip, which included continuing my travels to France, where I had been invited to give a course at summer school in the Pyrénées mountains. I was extremely depressed and could hardly imagine standing before an international audience and delivering my lectures while Soviet tanks entered Prague to suppress the ideas of the Prague Spring.

After a brief stay in Moscow in a state of desperation, I took my seat on the Moscow–Paris plane.

What great luck! Kapitsa was also flying to Paris. At that moment, when external events were developing in a catastrophic direction—the invasion of Czechoslovakia was a real national tragedy for Russia—and as every conscious man was developing a deep internal conflict, there was a special need for getting closer to each other.

Kapitsa, his wife, Anna, and I spent the whole day together in Paris. We have probably never been closer to each other than at that very moment of high tension and anxiety. Kapitsa was approaching his seventy-fifth birthday—a man who had survived two Russian revolutions and had long experience with living in both worlds. I wanted to rely on his wisdom.

"Nothing is terribly new," he said. "We should consider everything from a broader philosophical point of view."

Suddenly, his argument took a very strange turn. He said, "I'm not going to apologize before our Western colleagues at the Pugwash conference. Both sides, from time to time, do shameful things. Instead, I'm going, for the first time in my international life, to deliver a talk in pure Russian. I want to quote a famous Pushkin poem that our great poet published illegally[2] in a form of nineteenth-century samizdat."

2. Because official censorship classified the language of the poem as somewhat obscene.

Now, I can try to translate a small piece of Pushkin's words into English, a job Kapitsa didn't want to take upon himself: "In somebody else's eye, you would see even the straw, but in your own, you wouldn't notice even a log."

I have to confess that it is a slightly altered version of Pushkin's actual poetry, but the most important message for me was that Kapitsa was indeed too philosophical, and apparently too unwilling, to identify himself with the historic challenge to our national self-consciousness. At that moment, for the very first time in Soviet history, a new breed of people, younger than Kapitsa, represented the consciousness of the nation. These people were emerging as keepers of the flame of the free-spirited old Russian intelligentsia. It was the time of Sakharov and his friends.

Events in Czechoslovakia in 1968 created a strong political tidal wave, affecting the life of everyone in our country. It had not only precluded subsequent development of an active intellectual and social life, but also imposed the strict rules that came to characterize the stagnation of the Brezhnev era. No wonder it was dangerous to raise direct protest against the invasion of Czechoslovakia, and no one even tried to do so in Akademgorodok.

However, a group of courageous young scientists raised their voices, demanding reconsideration of an unjust governmental verdict against a few of Moscow's dissident writers. Two of my best pupils, Zaharov and Zaslavsky, signed that letter. In the post-Prague autumn of 1968, their gesture had been quite risky. The authorities in Akademgorodok put pressure on the institutes the cosignatories had come from. Budker was quite concerned that potential retaliation might damage the unique position of our institute. We all knew the insidious character of the regime that would not hesitate to hold all of us responsible for the dissent or perceived misbehavior of a few.

To minimize the risk, Budker even tried to persuade our young colleagues to remove their signatures from the undesirable letter, telling them the institute—such a bright spot in Siberia—should not be kept hostage in a confrontation with the government.

I, too, was under strong pressure as a mentor of the "nasty boys." I remember that at about that time we had of one of our regular gatherings at the institute's round table. Though somewhat depressed, we were still able to discuss the issues of everyday life and science without losing the physicist's ability for dark humor.

At that meeting I made the unforgivable mistake of proving that too much humor can be dangerous and counterproductive by itself. Budker told us that as a result of very successful negotiations with the government, he had secured the delivery of huge quantities of lead bricks, so badly needed for one of the future high-energy accelerators as a protection shield against radiation. However, authorities allegedly demanded that the huge carload of lead bricks be removed from the railway station within a couple days. So our director invited everyone in the institute to participate in a collective physical exercise. In an innocent joking mood, I suggested that we give the young dissenters a larger quota of bricks for unloading. Everyone laughed. Probably no one realized at that time that such a joke might be considered politically too sensitive and thus incorrect.

In the heated moral atmosphere of Akademgorodok, this story was picked up and eventually delivered to Moscow. I don't know in what form it was finally delivered, but at first Leontovich took it quite seriously. He invited me to Moscow for the express purpose of giving my explanation of the incident.

This general crisis in the life of the intellectual community coincided with my personal crisis. Suddenly I realized how high my own level of anxiety was. I perceived the framework of Akademgorodok to be more and more narrow. By 1970, I had to decide whether my stay in Siberia had indeed reached the saturation point.

Nearly a decade after the establishment of Akademogorodok, the mood of the scientists was very different than it had been in the early years. There was almost no infusion of fresh scientific blood. The shortage of apartments set up an obstacle for newcomers to join us. The influence of the local bureaucracies was steadily rising, while the pioneers of Novosibirsk, like Lavrentiev and Budker, were almost exhausted, both physically and emotionally. There was also a great deal of infighting within the establishment. That, and a decrease of social interaction and intellectual freedom, made our social life seem empty.

The success of our institute in science and in winning special governmental privileges complicated relationships with our siblings, the other institutes at Akademgorodok. It was not simply the issue of envy or jealousy: all of us had to live in one and the same environment and compete for a place in the apartment houses and in the recruitment of a qualified labor force.

But the argument of "the superiority of bright heads" wouldn't mollify anyone. In that sense the institute, almost from the very beginning, had been in a state of circle defense to protect its position and predestination.

The atmosphere was too drastic a change for me. Wasn't our academic city modeled after the Western university campuses, to serve as a haven of intellectual thought and freedom? The whole concept of an "academic city" had been introduced to create an impetus for the scientific community to advance more quickly. Indeed, in the early years of Akademgorodok, the creative scientific atmosphere was protected by the very notion that our community was self-contained—sealed off from the inhospitable environment of the day-to-day struggle for the bare necessities.

However, the artificial self-isolation of the academic community eventually backfired. Akademgorodok lost its most important asset, that of being in a state of continuous renovation and bringing in fresh blood. Every position in the institutes and all the apartments were occupied, leaving little chance for further change. The situation conflicted strongly with the nomadic nature of scientists as a singular species.

Once, back in the happier years of Novosibirsk, I had asked Budker what he thought the eventual fate of Akademgorodok would be—this island of science in a Siberian taiga.

"Andrei Mikhailovich," I had said, "I think it is counter to the second law of thermodynamics. How can one hope to keep this oasis of science and intelligent life protected from the everyday pressures whose tendency is to level, to equalize our resources with the shortages everywhere else in the country? It would be better to land on an island in the ocean, but even then you would need permanent, steady support and attention from the top levels in the government. Otherwise, it is an artificial, unstable configuration like in fusion plasma."

With an ironic smile on his face, Budker replied, "You think our system, as such, is more stable? Do you really think that this configuration, established the same year I was born, could find a smooth way out? Let me tell you a Jewish joke. A rabbi in a provincial town in the Ukraine was asked an abstract question: What would he do if the borders around the country were one day opened? 'I would climb the tallest tree,' he said.

"'But why?'

"'Very simple,' the rabbi said. 'So I wouldn't be swept away by the crowds rushing to leave.'"

On the professional side of my life, I became more and more involved in cooperation with Artsimovich and had less and less a direct tie to Budker's institute. The climax in our relationship was reached in 1970, when Budker suggested that I start a new institution in Akademgorodok that would reflect the loosening of my scientific contacts with his grand ideas about high-energy accelerators and his substantially more modest appetite for an engineering approach to controlled fusion.

But I had realized that the time for a fresh start in the stagnating atmosphere of Akademgorodok had passed. Finally, the painful decision was made. My family and I were ready to move back to Moscow. But I had only very vague ideas about what my future haven would be. The Kurchatov Institute was rejected mainly on psychological grounds. The past is past and should never be tried again. Also, I was determined to stay within nonclassified science.

The moment I identified myself as contemplating the return to Moscow, I became almost persona non grata in the official circles of Akademgorodok. I was not offended at all by such an attitude. After all, on a much smaller scale, it resembled the official reaction toward defectors and refuseniks.

I left my home of ten years with mixed feelings. Perhaps the concept of Akademgorodok was a premature one in the life of our country. It had been one of Khrushchev's attempts to salve the wounds inflicted into the system by Stalin's personality cult, and to make a giant leap forward to bring "this generation of Soviet people" to the communist future.

History will provide the final judgment on the success or failure of the Akademgorodok experiment. Possibly, some will conclude that it was a premature experiment that had been able to establish, and only for a while, something similar to the Utopian "City in the Sun." For those of my colleagues and friends who are still in Akademgorodok, witnessing declining support from the authorities and growing internal conflicts, there is still great hardship and struggle for the survival of science.

18 / Back to Moscow

➢

The choice of temporary asylum in Moscow was made rather quickly. In 1971, I was appointed head of the plasma theory lab at the Institute of High Temperatures. It was a young but rapidly developing institution—the grand vision of two ambitious professors from the Moscow Energy Institute, the same place where I had taught part-time before my departure to Novosibirsk. Since the early sixties, they had dreamed of building their own research institute to promote an innovative approach to the energy arena. They knew that as a prerequisite they would have to build their own power base.

One of the professors, Vladimir Kirillin, was eventually elected a member of the energy division of the Academy of Sciences. He had a talent for Soviet-style politics and management. After the death of Stalin in 1953, the upper echelons of the party hierarchy were reshuffled. Climbing the party hierarchy in the mid-fifties, Kirillin reached the key post of head of the science department in the Central Committee of the Communist Party of the Soviet Union (CPSU). The person in such a chair effectively played the role of politcommissar for the whole Soviet scientific community.

Kirillin, later to be my new guardian angel, already knew what was allowed and what was not for a supervisor of science. He never tried to intervene actively in the fate of Soviet biology, since Khrushchev had clearly indicated his intention to assume the role of Lysenko's top protector.

Kirillin and the Academy of Sciences suddenly accepted the notion that biology should not be considered a regular science—which, of course, was precisely true in the case of Lysenko-type biology. In every other respect, the position of head of the science department in the Central Committee of the CPSU represented a major power base. No new institute could be organized in the country without his approval. No director could be appointed to any of the scientific institutions without preliminary agreement from the science department, and guidelines for the elections of new academy members had to be approved by the polit-commissar. However, even more striking was the way the party hierarchy established absolute control over the foreign trips made by every scientist.

I first met academician Aleksandr Sheindlin, Kirillin's closest friend and the creative scientific leg of the tandem team they had formed to establish their new institute, in 1961 during one of my early foreign trips abroad. We Russians quickly adopted the point of view that exit permission to take foreign trips was almost a ticket to outer space. A feeling of great responsibility came out of our vigilance not to be seduced by the capitalist environment. This solidarity pushed companions on such trips closer to each other. It was clear for us that human nature was tested much more acutely on the front line in the trenches.

In that sense Sheindlin, already a mature professor, and I, a young scientist, naturally developed a mutual sympathy and respect. I held my companion in high regard for his genuine preference for the scientific agenda over sightseeing and shopping.

I suspect I also made an impression on Sheindlin—with my courage with the English language. Once, after midnight in a New York hotel where we were staying, my friend's room was suddenly flooded by flushing toilet water. To avoid a potentially dangerous international conflict, I was asked to be the negotiator. With the help of a dictionary, I found the English word *accident*. I'm pretty sure that my pronunciation was absolutely terrible when I talked on the phone, but the important message was delivered. Thus, ten years later my old friend academician Sheindlin came to *my* rescue—by providing temporary political asylum in the Institute of High Temperatures.

Kirillin, in the meantime, had moved to a new office as deputy prime minister to Alexei Kosygin, in which position Kirillin served as the government's supervisor of science and technology. Brezhnev, who was gradually strengthening his position as general secretary of the CPSU, appointed to the chair of the science department in the Central Committee his long-time protégé from the provincial party hierarchy in the Moldavian Soviet Socialist Republic. It was very difficult to identify the area of scientific expertise, if there was one, for this newcomer, Sergei

Trapeznikov—a man with an emotionless, cold face, whom I saw sitting in the presidiums at high-level academic gatherings. He rarely spoke, so I don't even remember his voice at those moments when he was delivering the party commandments for science. His scientific achievements were praised with vague expressions, naming him an outstanding expert on the history of Soviet collective farms.

Kirillin justifiably deserved the respect of the scientific community. Within the limits of the possible, and with the talent of a diplomat when necessary, he tried hard to help science, avoiding direct confrontation with the system. Probably his single biggest attempt to promote the science budget, in fact, was to create and help build the Institute of High Temperatures.

In my time, Kirillin was a very rare guest of the institute, though his name was imprinted on the plaque on the door of his symbolic office, next to the office of Director Sheindlin. It may well be that Kirillin never had any hands-on involvement in promoting the institute.[1] Maybe just the notion that "he is over there" was enough to support the skyrocketing budget and the scale of construction work under way at the institute.

The atmosphere at the Institute of High Temperatures reminded me of the one I had just left in Akademgorodok. Like Budker, Sheindlin promoted gatherings, sit-ins, and open discussions. He clearly preferred to rule by persuasion rather than by order. He liked to think aloud, and I believe a slight stammer greatly enhanced his enthusiasm for long speeches at official meetings—as well as equally long toasts at dinner parties.

Kirillin and Sheindlin wanted to use their talents and influence in the government for benign purposes. However, it was clear to me that we desperately needed a system based on objective assessment of proposed projects and ideas, and peer review by unbiased experts. This would have created an insurance and some safeguards against such one-sided, unjustified, inflated projects as the ones I witnessed at the Institute of High Temperatures.

At the time I joined the institute, Sheindlin entered into a new alliance on top of the one that he had with Kirillin. His new clients were from military industries. Fortunately, I had no relationship to his new

1. Purists would classify it as a conflict of interests.

passion—but word of mouth in the institute told us that the project in question was an attempt to generate high-powered microwave beams against flying rockets.

For several years, it was apparently the most expensive part of the institute's budget. Poor Peter Kapitsa! I believe he never knew that his extravagant fantasies of saving his home country from nuclear attack—invented by him twenty years earlier in solitude under house arrest and communicated in a letter to Comrade Stalin—would finally open a black hole that would truly suck in the taxpayers' money.

Thus, rather soon after joining the new institute, I developed a deep skepticism about how science at the institute was run. However, I was a guest provided with political asylum, and I was subject to a kind of extraterritoriality.

Sheindlin helped me bring Galeev from Novosibirsk and let me establish a small but quite solid group in plasma theory. We were absolutely free to choose our own topics for research.

The institute was run by what was called an academic council, a collective forum that confirmed the most important decisions on science policy and faculty appointments. It consisted of two distinct types of persons: two dozen of the leading scientists working in the institute whom I saw every day, and half a dozen prominent academicians working elsewhere. Their part in the academic council was rather symbolic—at least I never saw them at any of the meetings. Sheindlin probably placed their large portraits on the wall of the council room in lieu of their presence. It was indeed a most impressive collection of celebrities.

Showing the gallery to me, Sheindlin added, "Your portrait will soon be hanging next to these fellows."

I couldn't suppress an ironic remark: "That means I wouldn't have to hang around anymore!" In fact, it was a prophetic joke, even if unintended. Such a loose involvement in the everyday life of my new home institute gave me enormous freedom for extracurricular activities, the most important of which was the very close, intimate cooperation I had with Lev Artsimovich. By the time of my return to Moscow, Artsimovich had just had his sixtieth birthday—a milestone that is often considered in academic circles to signify the age of serenity and liberation from everyday boring routine. This age brings with it the moral right to focus on problems that almost touch eternity.

Indeed, Artsimovich concentrated more and more of his activities around the life of the temple of immortals, the Academy of Sciences. His sixtieth birthday was celebrated by us, his pupils and colleagues, with new rock operas, poems, and toasts. While still in Novosibirsk at Akademgorodok, I had designed, together with Zakharov, a special medal, which we called "In commemoration of the overthrow of Bohm diffusion." On the scientific side, it was to acknowledge Artsimovich's major contribution to experimental plasma physics and controlled fusion: the success of the Soviet-made tokamak.

However, it also had political implications, too. (Physicists never stop joking.) We demanded that the delivery of the award be accompanied with a loud recitation of the special certificate on how to treat the medal: "It should be fixed on the opposite side of the chest from the medal for the seizure of Prague."[2] Artsimovich, a real free spirit, especially liked this very important distinction, clearly congenial with his own political thinking.

By that time, it was widely known in the circles of the Moscow intelligentsia that Artsimovich had been one of the organizers and draftees of a famous collective letter sent in 1970 to General Secretary Brezhnev. The authors of this letter, prominent intellectuals and cultural figures,[3] warned the government of the danger of reviving the ghost of Stalin. This tendency, instigated by conservatives and reactionaries of the post-Khrushchev period, was clearly visible, even in the official documents of the party congresses. I remember the general assessment in our circles was that the letter had a certain positive impact.

Not long after my arrival in Moscow in the winter of 1971, Artsimovich invited me to a private dinner in his home, telling me in advance on the phone that he was going to appeal to my citizen's conscience. Intrigued by this, I soon joined a small company of physicists—among them Arkady Migdal and Sergei Kapitsa. To my surprise, I found among this group Bruno Pontecorvo, now complimented with the patronymic

2. This was meant in parody of the Soviet medals that were awarded after World War II for participation in the "liberation" of various Eastern European capitals. The instructions indicated that these medals should be worn on the "wrong side." Our medal was a pre-glasnost attempt to ridicule the then-recent actions of Soviet "liberation" forces in 1968.

3. Including, for example, the famous ballerina Maya Plissetskaya.

Maksimovich. He clearly felt quite accustomed to Soviet life, after the almost twenty years since his famous defection from the British nuclear establishment.

The purpose of the meeting was to consider the situation and draft another warning to the government. The reactionaries at this time in Soviet history were more subtle than to openly make hard-line statements in the official documents of the Communist party. Instead they launched an attack on the cultural front. Newspapers and literary magazines were flooded with articles and novels perpetuating neo-Stalinist slogans.

This campaign reached its culmination with the publication of a novel by a writer prominent among the hard-liners, Vsevolod Kochetov. Even the title of the book was highly provocative: *So, What Do You Want?* It was an open challenge, directed against the liberal intelligentsia and mixed with nostalgia for the totalitarian lifestyle and mentality.

We were concerned not so much by this published fiction, but by the encouraging reactions from party ideologists. Not without reason, Kochetov was considered an "official writer and spokesman."

I remember how we, the cosignatories of this letter, had a special objective: to make its delivery confidential. The previous letters of concern, sent to the government by different groups of liberal intelligentsia,[4] had immediately become known to the public at large. Usually, the very existence of the letter, and even its content, were communicated by one of the "enemy radios" like the BBC, Voice of America, or Radio Liberty. In the subsequent internal debate, often on the pages of the Soviet press, the authors of such letters were accused of being interested mainly in being promoted by the foreign mass media. Artsimovich and the rest of us came to the conclusion that no matter how the actual leakages had taken place, they preempted (or discredited) the delivery and impact on the letter's principal recipient, and in that way, they stole the effect.

We suspected it would be difficult to get our letter directly to the top. We had the feeling that someone was interested in undermining the final result. Contemporary analysts have already been able to trace such leakages of the manuscripts and even the videocassettes—of Solzhenitsyn and later of Sakharov—as something sponsored by the state security organs. However, the letter writers' wicked approach, entertained in order to alienate Brezhnev and the highest leadership of the liberal intelligentsia, ultimately served to get the information to wider audiences.

We felt a great deal of joy in actually drafting the letter. Indeed, the

4. Such letter writing was a familiar feature of the early stagnation period.

book in question gave us a lot of opportunities to ridicule Kochetov. In his novel, he portrayed the decadent mentality of the intelligentsia in sharp contrast to the party faithfulness and integrity of the veritable Bolshevik hero. In our letter, we offered a sarcastic remark about the hero's favorite moments, when he had secretly enjoyed looking through the window of the apartment across the street to see the regular striptease of the young lady before she turned out the lights.

Among other targets of our attack, I remember Kochetov's attempt to discredit Italian Communists. One of the negative heroes of the novel was a prominent Italian intellectual who supported the "revisionist" ideas of a sort later called Eurocommunists. The very fact that Bruno Pontecorvo had been invited to join us was explained by the Italian connections in the novel.

The one thing that was not included in the final letter was Artsimovich's remark: "Now everyone can understand why we need the pro-abortion law. It would provide one more opportunity to prevent the birth of the likes of Kochetov." We thought that was perhaps too much.

After we had completed our letter, we sent it. The letter was signed by about a dozen different people from our circle. As a rule, those who were invited to join us expressed great enthusiasm. Everyone understood that the Kochetov story was only an excuse, used as a pretext for one more warning of a return to hard-line policies.

There was only one refusal of Artsimovich's invitation: Peter Kapitsa. We were, of course, quite disappointed. Few people knew at that time how much Kapitsa himself was expressing in his own letter writing. What we considered an outstanding display of our civil courage had been for Kapitsa something quite elementary and routine for many years.

The real courage, however, in the fight against the totalitarian regime and mentality was shown by a small group of other people: Andrei Sakharov, Yuri Orlov, and their friends. While we had a rather similar assessment of the totalitarian and later autocratic rule of the Communist party, the ways in which we were ready to express our convictions were quite different from those of Sakharov and Orlov.

Artsimovich and I, along with our like-minded moderates, were in favor of acting within the system, avoiding the alienation of the government and authorities. Maybe subconsciously we simply were not ready to sacrifice the privileges given to us by the system—the chance to do science and enjoy certain well-controlled dosages of foreign trips. We thought that the expression of our concern in letters or in speeches at academy gatherings would, in the final account, be able to produce the net effect. The authorities played against our moderate internal opposition with the carrot and stick—and they knew that we didn't much like the latter.

But Sakharov and his friends were not afraid of the stick. They had already "burned their bridges" to the system and openly declared disobedience—and even moral civil war. I have always had admiration for people of such heroism and self-sacrifice. However, most of us at that time considered them Don Quixotes.

19 / The Academy of Sciences: The House of Lords

➢

For members of both groups—open dissidents or those who espoused moderate internal opposition—the status of membership in the Academy of Sciences provided a kind of protection. For a number of years, until his final exile to Gorky, Sakharov was considered untouchable—in the sense of direct physical threats—while most of his colleagues, "mere mortals" from the human rights movement, were arrested and sent to prisons or psychiatric asylums. The independent stance of Kapitsa and Artsimovich had certainly brought to them more respect from the scientific community and a feeling of displeasure from the authorities. In an odd way, the status of academician, introduced in 1726 as a postmortem implementation of Peter the Great's will, provided a certain degree of political protection throughout the Soviet era.

Artsimovich, prominent in academy affairs, appointed me his deputy in the Department of General Physics and Astronomy. He gave me the chance to learn about this unusual animal called the Academy of Sciences, a relic of Soviet society from distant prerevolutionary times.

Created in order to bring the seeds of European culture and knowledge to Russia, the academy was originally built on brains drawn from the European centers of civilization. For several generations, science was at the court of the czars, and one of its duties was to glorify the imperial crown.

In the beginning, this glorification was successfully done by such foreign trophies as Leonhard Euler, the greatest mathematician of the eighteenth century, or Daniel Bernoulli, the founder of fluid dynamics. Later, leadership was taken by such native Russians as Mikhail Lomonosov in the mid-eighteenth century, and a hundred years later, Dmitri Mendeleev, author of the periodic table of the elements.

From the very beginning, the Communist government had a strong antipathy to the very idea of the Academy of Sciences, a creation of old Imperial Russia. According to the original scenario implemented by Lenin, the whole social layer of the old intelligentsia had to be replaced by the new Soviet man. The academy must be tolerated only to a certain degree. In parallel with it, the government created the Communist Academy, which they envisioned would take over in the future. "The right scholars" were promoted to this body (like Konstantin Tsiolkovsky, for example). However, the government quickly learned that the party's loyalty was not sufficient to produce practical scientific results, and the Communist Academy, dominated and even polluted by Marxist philosophy, was dissolved in the mid-thirties.

The leadership adopted another tactic. It decided to gradually infiltrate the Academy of Sciences with new men. The first massive transfusion of "fresh blood" took place in 1929 under very strong pressure. The academy was forced to elect almost half of its new members from a list of influential party bosses suggested by the government. The party thought that the best way of providing at least a minimum of compatibility in the academy would be to send Bolsheviks who had cultural roots in the old intelligentsia. This criteria was later to backfire, when many of them, like Nikolai Bukharin, became the victims of Stalin's purges.

By the time I came to the academy's headquarters in the seventies, the interactive process of regular transfusions and unspoken silent concessions on both sides had brought the composition of the academy to a ratio in which about 80 percent of the academicians were Communist party members. The Academy of Sciences was no longer in confrontation with the system.

Still, it was a bizarre mix. It was an old prehistoric structure with privileges and party leverages to keep it under control. Membership in the academy was based on a kind of silent convention to share the seats between genuine scientists and protégés of the authorities. Elections were based on multistaged secret ballots with all of the attributes of visible democracy.

However, the government had a lot of buttons to press. When it was clear that brute force would not work, special vacant chairs were created in order to promote government appointees and protégés. In such cases, a secret ballot was reduced to simply a no-confidence vote. Such a de facto balance of power yielded the coexistence of "good guys" and "bad guys."

The status of membership in the academy provided a number of privileges and even a decent lifelong stipend. Five hundred rubles a month for a full academician and 250 rubles a month for corresponding members was a huge amount of money in the seventies. At that time, membership effectively meant the doubling of one's net income.

With the combination of prestige and respectability, election to the academy became attractive in the eyes of the party elite. Academic regalia became a real magnet. Stalin was decorated with the title Honorary Academician.

And by the early Khrushchev era, the only person who had inherited such a title from the epoch of the personality cult was Molotov. The academy found an elegant way to eliminate this open rival to Khrushchev from its ranks. It simply canceled the category of honorary membership at once. Indeed, it was a very wise decision. Otherwise, Khrushchev himself, being elected to such a rank, would have created another problem for the academy later, after Brezhnev's coup d'état.

In a certain sense, the Academy of Sciences operated like the House of Lords. The official gatherings of the 200 academicians had the right to adopt bylaws and decisions defining the life of the huge scientific community in the country. Yet, since the academicians, by the academy's charter, did not represent the broad scientific community, the tens of thousands of scientists and employees working in hundreds of different institutes effectively had no voice in academy affairs. In that sense, the ordinary scientific worker was a kind of intellectual serf to the academic barons who reported only to themselves.

For the government, the most important task in its control over the academy was to make sure it was their man who occupied the position of president. It had always been a trade-off between party loyalty and scientific credentials.

The appointment of Mstislav Keldysh to the post of academy president marked the advent of new times. The rather inoffensive "platonic" scientists of the past gave way to the descendants of the military-industrial complex, of which Keldysh was probably the most prominent representative.

As a matter of fact, he had a very good record in the academy as an expert in his own scientific discipline. When he was young, he started as a bright mathematician and became a full member of the academy at the age of thirty-five. Gradually he turned his attention to applied issues in an attempt to help solve such formidable problems as flutter for airplane designers. In the late thirties and early forties, this phenomenon of vibrations in airplane bodies, coupled with the surrounding air streams, was a principal obstacle to the progress of aviation.

Keldysh spent several years working in various mailboxes before he finally became director of the Institute of Applied Mathematics, the leading place that developed computational techniques toward strategic applications, such as the nuclear and rocketry programs. His particular contribution was especially important to the space program. Here, he became second only to Korolev and was in charge of all the orbital dynamics of spacecraft.

Thus, his promotion to president was a first in the history of our science: the marriage of the academy as the headquarters of pure, basic science to the military-industrial complex.

I remember an offensive joke that circulated among the scientists at that time: "Instead of representing the academy in the Central Committee, Keldysh represents the Central Committee in the Academy of Sciences."

Artsimovich introduced me to Keldysh, and later I myself had countless occasions to spend hours in discussion with him, sometimes even bitterly debating different issues of space science and its organization. Clearly, he was an outstanding man, who combined the unusual qualities of a bright scientist and a man experienced with the applied work at the heart of the military-industrial complex. However, I believe such a combination always evokes a kind of deep internal conflict in a person. Sometimes I witnessed him genuinely suffer from an almost physical pain brought on by his conflict between Dr. Jekyll the scientist, and Mr. Hyde the administrator, the hostage to the system.

Keldysh tried to launch a package of progressive reforms in the academy. One of the most urgent among them was his suggestion to rejuvenate the old boys' club. Not only was the average age of the academicians becoming older and older, but even the very age when a person was introduced to academy membership was slowly increasing. But the aging of the academicians produced two parallel tendencies. The physical deterioration and the psychological transformation of aging eventually led to a deterioration of scientific standards. The older the academician, the more that person became detached from the younger and more creative scientific community, and stuck with like-minded contemporaries. Any academy that operates as a club of mutual admiration can fall into such a trap.

With the help of the government, Keldysh literally forced the academy to impose an age limit for the election of new members. While the electorate had obviously opened the academy's doors to the younger generation, it gave preference to candidates very close to the threshold. Poor fellows! This was their last chance.

I personally should not complain at all. I was elected a corresponding member at the age of thirty-two, and four years later, a full academician.

The division of responsibilities between different departments of the academy was based on their topical principles, and we introduced our own classification of different scientific disciplines in the academy: the natural sciences, which was us; the unnatural sciences, which was the social sciences and the humanities; the counternatural sciences, in which category I would place all the weapons designing; and the supernatural sciences, consisting of nonstop ideological sermons and Lysenko-type pseudoscience.

The highest gathering in the academy—its presidium chaired by Keldysh—would have been a one-man show had it not been for the presence of Artsimovich and Kapitsa. Only these two academicians were capable of challenging the domination of the president. They challenged Keldysh's attempts to reform the academy into a part of the governmental bureaucracy. Keldysh slowly but steadily promoted a policy to convert the academy into the formal headquarters of all technological and scientific development in the country. Artsimovich was the most eloquent opponent of these ideas, feeling that the temple of pure and fundamental research could easily become the Ministry of Science.

Keldysh argued: "It's not me who is a bureaucrat; it's you, Lev Andreevich, who is the retrograde, trying to confine us in the Monastery of Science."[1]

The Academy of Sciences was given responsibility for final judgment on any nationwide initiative, whether it was investment in a new branch of industry or a construction project that might have environmental repercussions. Even official party documents and speeches of

1. Despite this conflict, outside of academic games and battles they were good friends who respected each other. I was present at a few warm home-style vodka parties dominated by these strongmen—strong both in the consumption of spirits and in the very high spirits of their discussions. Sometimes, the discussions finished with friendly teasing between the two of them about whose pedigree was more noble. Despite the fact that Artsimovich's ancestors were of higher nobility than those of Keldysh, in academy affairs, Keldysh was above Artsimovich.

Brezhnev reflected this important transformation. Science, it was declared, "is becoming a productive force in society." What a consolation for the intelligentsia, for the scientists! Finally, we might acquire the same civil rights as the ruling working class.

The academy was, unfortunately, not always able to properly use its role as national referee for a good cause. The political and corporate pressure from interest groups quite often forced the academy to provide a rubber stamp for a number of national initiatives later condemned as environmentally or otherwise dangerous. Only fifteen years later, the decision to build the paper industry on Lake Baikal was reversed, and the academy had to accept that they had sinned.

On the other hand, the academy was never asked about its attitude toward Lysenko-type biology, and it always tried as much as possible to defend the geneticists. Aleksandr Nesmeyanov, when he was still president of the academy in the fifties, started his own alternative approach to the resolution of the national food problem. Agricultural production, controlled by the party and Lysenko, was in a state of complete decay. So Nesmeyanov suggested a purely chemical way to feed the population. He argued that contemporary organic chemistry could, in a long chain of reactions, transform fossil hydrocarbons into a protein-rich food.

I remember vividly the scientific session of the presidium, chaired by Keldysh, when Nesmeyanov delivered his concept. His team of young, good-looking female assistants distributed the edible products of such experiments. Nesmeyanov explained that up to that point his institute had made only limited progress. Nevertheless, he demonstrated synthetic steak and synthetic black caviar derived from real milk, not yet from coal or petroleum. A faithful patriot, he clearly tried very hard. After all, it's not a tragedy that he did not succeed with that type of chemistry.

However, Khrushchev and then Brezhnev borrowed the principal concept of the transformation of fossil hydrocarbons, but in a slightly altered form. First they turned the oil and gas into hard currency, and then they purchased Western grain. Such a conversion helped prolong the decay and agony of a political and economic system unable to feed itself without substantial grain deliveries from non-Socialist parts of the world.

20 / The Beginning of a Space Career

➢

Artsimovich, whose friendship and cooperation became indispensable for me, especially upon my return to Moscow, suddenly died of a heart attack in early 1973. He was a strategic genius in finding the most efficient and economic solution in carrying out the scientific research of his controlled fusion team. He was also a master at administering and supervising his colleagues in his division of the academy. That plus his rare political instinct would have been absolutely priceless during the turbulent years of perestroika. I often wonder how he might have reacted to current events, after the more than seventy years of one-party domination that had force-fed the population with its compulsory ideological rations. Artsimovich always made clear and ruthless assessments of the system's follies.

In those days, after the loss of Artsimovich, I felt quite alone, almost suspended, in an undefined realm of my own. I knew that the Institute of High Temperatures was going to be only a temporary asylum, but I was not yet ready to make up my mind about what to do next.

Without much enthusiasm and maybe only by inertia, I continued the modest administrative job in the academy that I had taken to help Artsimovich. Several times I was asked by academicians President Keldysh or Vice-President Vladimir Kotelnikov, a physicist and Artsimovich's colleague, to brief them on events at different institutes in the Department of General Physics and Astronomy. Sometimes they asked me to convey specific instructions to be implemented. I felt some sympathy and interest toward these two academicians. Probably in a certain nostalgic or sentimental way they considered me the keeper of the Artsimovich legacy.

With some sadness, and at the same time not without a smile, I often recall the way Artsimovich introduced me to his friends in the Academy: "This man is the most famous Tatar after Genghis Khan!"

I would decisively dismiss such a genealogical line.

"Lev Andreevich! Once again I must issue an official protest! The

Tatars on the Volga had nothing to do with the Mongol hordes led by Genghis Khan or Batu Khan. My ancestors were ethnically, culturally, and even religiously completely different from the Mongols. As a matter of fact, Tatars suffered from the Mongolian invasion even more than the Russians did! Their capital, on the Volga, was completely leveled and burned twice in history, and it was never able to recover. Even the very name *Tatars* was given to us by the Russian czarist empire."

Then Artsimovich, surrendering, would interject with a friendly reply, "Well, Rolik, if you aren't the descendant of Genghis Khan, I will suggest another line." And he would turn to his colleagues and continue.

"My friend Rolik is a veritable pacifist and prefers to be introduced as the most famous Tatar after Ulugbek."

By that time, my fervor in resisting him would be gone. After all, Ulugbek, the grandson of Tamerlane, the emperor of Samarkand, spoke a language of a Turkish group, which I probably would have been able to understand.

At the age of fifteen, Prince Ulugbek had succeeded as ruler of one of the greatest empires of the fifteenth century, whose power extended over the huge areas encompassing the central Asian parts of the former Soviet Union, Afghanistan, Pakistan, and even a part of India. The young boy couldn't change the militaristic lifestyle of the epoch or the ardor of the royal army who had inherited his famous grandfather's spectacular military conquest. Furthermore, Ulugbek was even unable to maintain his kingdom's gains. His army lost one battle after another, and his huge and victorious empire started to shrink.

A wise man—an astrologer—told Ulugbek that indeed he had suffered a serious reversal of fortune, independent of his military art or the mightiness of his troops. Luck had left him because his stars were wrongly aligned in the sky at that moment in time.

Ambitious Ulugbek wanted to verify for himself all that had been said, and he wanted to add to his crown the masterminds of the celestial constellations. He launched an unprecedented search for the wisest astronomers in the medieval Orient. He brought them to Samarkand and built an impressive observatory. Step by step, the sky slowly unveiled its secrets before the emperor who was thirsty for scientific knowledge. Ulugbek was fascinated with the beauty of the stellar universe. However, his armies, and his subjects, discovered one day that the emperor was no longer interested in waging and winning wars. It was science—astronomy—that had won the heart and mind of Ulugbek.

Tourists to Samarkand can still see the remnants of the monumental observatory erected by Ulugbek. In its time, it was the biggest and most sophisticated astronomical instrument on the planet, the early precursor to the modern-day Keck telescope in Hawaii. The map of the sky, produced by Ulugbek and his astronomers, was remarkably accurate, and it included the positions of about 500 stars. If it had been available for sea travelers, it would have been indispensable for navigation.

Captivated by the secrets of the sky, Ulugbek lost touch with reality and was unable to navigate himself on earth. He fell victim to a coup d'état, in which he was not only overthrown but decapitated, too. His story is a ruthless word of warning to those who naively believe that one can combine the duty of a big administrator with a passion to do science.

There are different roads that lead a person to become a hostage to the mysteries of the universe. Mine was the least predictable or expected. In the spring of 1973, when I was still sailing in Moscow academic life without my own navigational star, Keldysh asked me to take the directorship of the Space Research Institute. It had only recently been established, with a goal to represent the interests of the Academy of Sciences in the national space program. I had had no idea that my future would eventually be intimately associated with the institution. However, I remember during our meetings the severe and merciless criticism of the institute's leadership. It came from the group of eloquent and outspoken members of the commission that represented the entire space industry, the Ministry of General Machine Building.[1]

The critics from the commission were mainly chief designers of different mailboxes, who operated behind the curtain of secrecy that protected the aerospace industry. Without being emotionally involved myself, I did not try very hard to understand the details of the conflict. The only message I remembered was that something was fundamentally wrong with the organization and management of the Soviet space program.

At first glance, the home country of *Sputnik I* and of Yuri Gagarin was busily engaged in launching an overkill of spacecraft into space—in

1. In reality, these so-called "general machines" were nothing more than launchers and sputniks. (The Ministry of General Machine Building was a separate enterprise from the Ministry of Medium Machine Building, with which I had been associated during my fusion years at the Kurchatov Institute.)

every possible direction—some of them flying to the moon, to Venus, to Mars. . . . The family of cosmonauts, the Soviet version of the "right stuff," was growing faster than in the United States. So what was wrong with the space program?

After Keldysh's offer, I had to do a great deal of homework about the program. My final decision, I thought, should be based on my own overall assessment of what was going on in the Soviet space program and at that particular institute. I asked Keldysh at our very first meeting only one question: "How much time can I have to make up my mind?"

He answered, "I think one week should be enough for you." Such a response was a clear hint that it was an offer that I should not even try to refuse. There were only a very few academicians who did not already have administrative responsibility. Of those that remained, even fewer were physically capable of doing such a job. I knew that the ultimate pressure to force me to accept the offer might be overwhelming. Still, I wanted to learn as much as I could about this terra incognita of my future.

How much did I know about space since *Sputnik I* was launched in October 1957? The Soviet press was literally obsessed with everything that happened in orbit—including elaborate descriptions of the cosmonauts' menu at their last breakfast and all the details of their physical exercise program. Every launch brought one, two, or even three more Heroes of the Soviet Union, and even more photographs of our space superstars, embraced first by Khrushchev, then by Brezhnev. The relentless propaganda spoiled the very idea of man's flying in space. The attitude toward our space achievements demonstrated, as we were told by the official press, the superiority of the socialist system.

For people on earth—at least the nonconformists on the street—these feelings were best expressed by the famous samizdat poet-singer of the epoch, Aleksandr Galich. His songs, which became popular in the early sixties, had lyrics like: "And we are busy making rockets and crossing the Yenisei River with a dam. And also in the area of ballet, we are ahead of all the planet."

My memories of the childhood excitement I had felt, stirred by stories about the conquests of Arctic landings on icebergs, were of a different time and place. The epoch of great geographic studies of the earth were long over. There were no more poles on our planet left unvisited, though there were countless worlds in space. But by that time, I was already mature enough to accept the notion that intellectual adventures in science, especially supported by the almost unlimited mental power of theoretical physics, was incomparably more exciting.

My perception of the scientific data brought back after the first spacecraft were launched was reminiscent of the great rush to dig gold in the Klondike or Siberia. It sounded more like a gamble than serious science, which deserved a solid approach.

I recalled that unfortunate visit of Vasily Mishin, the chief designer who succeeded Korolev at our Institute of Nuclear Physics in Novosibirsk, reeking of alcohol while he discussed Soviet space defense. Would I be working with this man or, even worse, would I be dependent on him? The failure of the Soviet lunar program and the unsuccessful attempt to build a superbooster to send the Soviets to the moon were, by that time, already part of the spoken samizdat.

My friends from Moscow State University who were associated with the space program told me stories that made me even more pessimistic. The guiding philosophy behind Soviet space launches reflected the interests of the space industry to the complete neglect of science per se. This was, no doubt, because the original motivation to build rockets had been purely military.

After the first explosions of American atomic bombs in the summer of 1945, the Soviet leadership under Stalin was paranoid over the perceived vulnerability of our country to nuclear attack. The United States not only had nuclear weapons and heavy bombers, but its bases almost encircled the Soviet Union—in Europe, in Turkey, in Japan.

The geostrategic configuration of our country did not favor a symmetrical response to building similar types of bombers. That is why the best minds in Soviet rocket technology were sent, immediately after World War II, to Penemunde, Germany, to gain what they could from the former headquarters of Wernher von Braun, which had been seized by Soviet troops.

By the early fifties, the young Soviet aerospace industry was ambitiously ready to launch the first intercontinental ballistic missile (ICBM). Strategically, if successful, it would close the window of vulnerability. The Soviet nuclear bomb had already been tested in late 1949, and Sakharov's team was actively engaged in developing the ultimate weapon, the hydrogen bomb.

Korolev and his colleagues could have only had a vague idea of how heavy the final reentry vehicle should be. Sakharov was still far from knowing how to make this deadly weapon relatively compact and easily portable. Since rapid progress was required, rocket designers adopted a worst-case strategy and started to develop an ICBM that, as it was discovered later, had a substantial excess of launch capability, or throw weight.

The biggest liquid-propellant rocket of the fifties was almost ready when Korolev, trying to go beyond the narrow and well-specified horizons of his military contract, discovered there was a great deal of interest among scientists in launching an artificial satellite to orbit the earth. After the successful test of the R-4 rocket, the way was opened for straightforward modification of the final stage, and Korolev persuaded the government to let him build and launch the first man-made object ever to orbit the earth.

The biggest international scientific event of the late fifties was quickly approaching. Known as "The International Geophysical Year of 1957–1958," there were thousands of scientists—geophysicists, geologists, oceanographers, atmospheric scientists—striving for coordinated measurements to greatly extend the knowledge of our own terrestrial environment. One could hardly suggest a better framework for the launch of an artificial satellite.

Sputnik I's launch into orbit on October 4, 1957, did not carry scientific experiments. There was not a single scientific instrument ready for the flight. Nevertheless, the impact of *Sputnik I* was enormous. It exceeded all of Korolev's expectations, not to mention those of the Soviet government.

While the Russian press had announced the news about the launch of *Sputnik I* in a rather routine and businesslike fashion, the world was surprised and shocked. It was a real eye-opener for the authorities, and maybe even for Korolev himself. Only after overwhelming international reaction did newspapers in the Soviet Union use the biggest print letters to characterize the outstanding success.

On that first flight, scientists had no dedicated scientific instrumentation, and there would have been no practical scientific output had it not been for the indirect data on the gradual evolution of *Sputnik I*'s orbit due to atmospheric drag force. This phenomenon helped to reconstruct the atmospheric environment at *Sputnik I*'s altitude, which was about 300 kilometers. The whole of the earth's outer atmosphere became an object of scientific research.

The Academy of Sciences immediately started lobbying the government and the space industry for the opportunity to develop dedicated scientific instrumentation to be launched into space. From a contemporary perspective, those early instruments, measuring atmospheric properties or plasma particles in the terrestrial ionosphere, were extremely simplistic. While the rocket designers had to learn how to design, build, and launch the spacecraft and satellites, scientists had to learn how to do science in space.

After *Sputnik I*, the authorities' appetite for launches and for the quick political dividends they brought was growing. But there was neither time nor place for science on *Sputnik II* either. Nevertheless, Korolev firmly promised that a third sputnik would be dedicated to scientific experiments. This came, many thought, not a moment too soon. My colleagues had felt especially pressed by the early American space program. Despite initial setbacks, the American program was slowly acquiring momentum. Scientific competition in space, on top of the space race, was imminent.

An early example of this pressure happened in the spring of 1958. A scientific team landed at the Baikonur Cosmodrome for the final integration and testing of hardware on *Sputnik III*. Korolev invited everyone for the last briefing before the final okay was to be given and the countdown started. He was pleased with the reports of the different experimental groups. It was the first impressive collection of scientific instruments, each of which was reported to be functioning normally.

However, trouble was soon discovered in some of the supporting hardware. The problem was with the more or less routine tape recorder, whose function was to accumulate data from different experiments and to prepare messages for the ground station. The spacecraft, revolving around the globe, would only be in contact with the ground station during periods of "direct radio visibility." Simply speaking, the ground station would be unable to sense the signals from the spacecraft when it was behind the horizon.

This tape recorder was placed on board to record everything the ground station would miss. With such a crucial role, members of the scientific team were extremely worried about the troubled tape recorder, and they recommended postponing the actual launch to give the technicians a chance to fix it. However, the tape recorder's ambitious engineer,

Alexei Bogomolov,[2] did not want to be considered a loser in the company of winners. He suggested that the testing failure was simply caused by electromagnetic interference from the multiplicity of different electrical circuits in the test room. He boldly proposed to launch *Sputnik III* on time.

To the great disappointment of the scientific team, Korolev accepted Bogomolov's suggestion, and final preparations were made for the launch. During the flight, however, it was confirmed that Bogomolov had been dead wrong. His tape recorder did not work. Consequently, the scientific information gathered was limited by the area of direct radio visibility, an area of about 1000 kilometers. Each scientific group had results, but because of the recorder failure they had to guess whether the phenomena discovered were of local or planetary significance. Especially intriguing data were collected by the group from Moscow State University led by Sergei Vernov, a distinguished Soviet cosmic ray physicist.[3]

Vernov and his coworkers had been the first to encounter cosmic ray particles in their natural environment. This is why they were awaiting the results from space in a state of utmost anxiety. Detectors were sending signals indicating extreme levels of enhanced radiation. What was it? A spot in space, somehow coincidentally located just above the ground station, or something more extensive? Did it even embrace the whole planet at altitudes above the atmosphere? Again, that damned uncertainty!

Without the verification of the tape recorder, a new experiment was needed and a new launch was required that would be capable of detecting radiation in any part of its orbit around the earth. Korolev promised to do it in one of the next launches.

In fact, only a few weeks later in 1958, a cosmic ray detector was launched to scan every bit of the satellite's orbit, but to everyone's dismay it was an American launch, not a Soviet one. It brought one of the most interesting discoveries of the early space era: the discovery of radiation belts around the earth. They are now called after James Van Allen, an American space physicist from the University of Iowa, who had

2. In my space time many years later, I too often had to depend on his hardware.

3. The whole science of high-energy physics in its early age was based on energetic particles reaching the surface of our planet. The atmosphere of the planet acts as a shield protecting us from deadly extraterrestrial radiation. This is why every experiment designed to study cosmic rays at the altitudes of artificial satellites, and later in interplanetary space, was considered a genuine chance to learn more about the primordial properties of cosmic rays.

successfully flown an energetic-particles experiment on an American satellite.

The Moscow team was left in a state of complete despair. The most they could do after the Van Allen discovery was to send their instrumentation on different satellites and spacecraft to confirm Van Allen's data and to gather additional details. But nothing conceptually new was discovered. Korolev, of course, was all too familiar with this dramatic change of score in cosmic ray physics in favor of the Americans.

A few years later, when he met Konstantin Gringaus,[4] Korolev himself raised this topic.

"You think that I don't feel guilty about that whole story? Korolev said to Gringaus. "You think I was simply a fool that day, issuing instructions to launch when the tape recorder had failed to work? If you want to know what happened, I can tell you.

"On that very unfortunate morning, Nikita Sergeyevich called me on the phone at the Baikonur Cosmodrome. He said that the Italian Communists had urged him to do something spectacular, like sending something into space. The next day, Italy was going to have parliamentary elections. 'If you Soviets would make one more show in space,' Khrushchev had been told, 'it will bring our Communist party a few more million votes.'"

The priorities were thus established. No one at the top had been serious about finding out what was wrong with the scientific equipment on board.

For me, the director-to-be of the Space Research Institute, the moral of this story was to stay calm and not to expect miracles for scientists in space. Science was only a hostage to high-level politics, of the games played by the government and party leaders in the corridors of power. At least Korolev had tried to be an ally of space science. But for me, the main issue would be to find other such potential allies. With Korolev gone, were we left with only Mishin?

My friends told me there was one other outstanding space engineer, Georgy Babakin. His ingenious designs of unmanned robotic probes to the moon were known as Lunnik and Lunokhod, or "moon rover." These toys, tiny compared to the size of *Apollo*'s modules, saved the

4. One of the members of the science team of *Sputnik III* and a colleague of mine from the later years of the space program.

prestige and face of the Soviet space program after the failure of Mishin's superbooster to send a man to the moon. Lunniks flew around the moon and photographed the hidden, invisible side of its surface. They landed on the moon and were roving under remote control from the ground station on Earth. They dug lunar soil, picked up samples, and brought them back to Earth with return rockets.

Babakin was contemplating even more spectacular unmanned flights and landings on Mars and Venus when I met him on the eve of my own space career. He was full of exciting, innovative ideas and receptive to suggestions from scientists. But he died before I was able to take up my future office at the Space Research Institute in the early seventies.

The final inventory of the facts I had discovered and their interpretation relevant to my future space career was rather discouraging. Many times during the week that Keldysh had given me to make up my mind, I was close to rushing in and telling him, "No, I'm not going to take your offer." But something inside stopped me short of making such a move.

In a rather philosophical mood, I came to the conclusion that scientists are indeed hostages to politics. I couldn't possibly refuse the offer. The only thing I could do was to follow Budker's advice: "When we cannot change the climate in general, let's concentrate on creating a better local environment." With such feelings, I went to say yes to Keldysh.

21 / Circle Defense for the Space Research Institute

➢

Upon accepting the role of director at the Space Research Institute, I had no illusions about the moral climate inside the institution. Not one of my friends who briefed me about my future "estate" painted an idyllic picture of universal friendship and love. Scientific institutions are generally very complicated organisms. The natural tendencies of different researchers or groups to challenge anything suggested by competitors, or even rivals, immediately creates an atmosphere of confrontation. In a normal institute, a few strong, bright leaders can usually redirect any excess energy away from potential confrontation to fruitful, if not friendly, nonexplosive competition.

My institute—IKI in its Russian acronym—was quite different. The barricades separating two conflicting camps lay not between the bureaucracies in the government, but between the space industry and the space science community. Deep divisions existed within the institute, which was split into a number of small strongholds—in the hands of different scientific clans.

I remember well Artsimovich's definition of IKI: "It's not a scientific institute; it's a travel agency for space science."

Indeed, between different laboratories and divisions of the institute there was little scientific common ground capable of bringing them together and creating one harmonious collective. The astronomers, dreaming of launching sophisticated telescopes to unveil "deep space," did not hide their disregard for the space scientists who studied the upper atmosphere and plasma within the vicinity of planet Earth, or even the solar wind. Despite the fact that early successes in space science were associated with this very discipline,[1] the leader of IKI's space astronomers, an outstanding and unusual character, Iosif Shklovsky, was sure that local "environmental" science would be nothing but a "*caliph* for an hour."[2] He was probably right on an astronomical time scale, but it

1. Such as the discovery of the Van Allen belt.
2. *Caliph* was the title of the ruler of Baghdad.

didn't help to create an acceptable psychological environment in the institute.

On the other hand, I thought that those who were involved in the early experiments, measuring the plasmas and energetic particles in immediate outer space, were intoxicated by their desires to launch their instruments on every spacecraft. The data files from space were accumulating at a tremendous speed, but the scientists had very little time to digest them. However, they continued to log more and more launches.

In the third corner, there was a group that comprise a new breed of space scientist. They anticipated that their instruments would soon be delivered to the surface of planets, with all the benefits and dividends reminiscent of the epoch of great geographic discoveries. They had already brought back a lot of data from the surface of the moon, and they had indirectly enjoyed the first contact with the atmospheres of Mars and Venus.

There were also scattered groups of geologists, both extraterrestrial and earthbound. These people were preparing for the opening of a huge program of remote sensing of our own planet.

In addition to the interdisciplinary barriers separating different scientific groups, there were old animosities and even open fights, which had been inherited from earlier times. All of the members of the institute had joined IKI during a short period of a few months after it had been hurriedly established by governmental decree. The main principles endorsed by Kapitsa and Budker—on "the selection of the cadre" and the importance of making appointments on an individual basis—were completely overlooked.

IKI, because of its random hiring, had become a so-called complex institute, where a great variety of disciplines were gathered under one umbrella. The academy had a number of such institutes, predominantly at distant locations.

Academy jokers immediately invented a formula that stated that "the complex institute signifies that its budget is its real part and its scientific output is its imaginary part."

While the scientific administrative division inside the institute was like a complicated pluralistic animal, party life per se in the institute, promoted by its Communist party organization and comprising about 20 percent of IKI's employees, was like a monolith.

I learned such details very soon from Keldysh himself. He told me, with an ironic smile, that he had had a request from the Moscow city party authorities to consider a letter of protest, signed by a group of leading staff members of IKI, on behalf of the Communist party bureau. The letter warned the authorities that the appointment of me—a non-party member—as the new director would bring the institute back to the bad experience it had had with the first director, my predecessor. However, Keldysh firmly dismissed the protest. The man whom I had to succeed, the former director of IKI, academician Georgy Petrov, in fact was probably the brightest spot in the institute.

Six feet tall and lean, Petrov had a shock of thick white hair. He was always smoking and simultaneously telling stories in infinite number. This assured that his interlocutors would have great difficulty in trying to insert their own argument or message. His friendliness and natural generosity, however, helped me a great deal during my first months in office. Petrov had been essentially fired from the job of director for being unable to manage the institute. But despite all the initial expected uneasiness in our personal relationship, we developed a very open friendship.

On one of my first meetings with Petrov after I assumed this post, he generously painted for me a huge canvas of the Soviet space hierarchy so I would know what to expect. He knew most of the protagonists of the space program and rocketry, and he told me that in the country's huge and complicated space infrastructure, he identified IKI's role as that of a sacrificial lamb. The big bosses of the space industry and the gray cardinals of governmental politics controlled the situation, and any hint of potential failure would be explained as the wrongdoing of IKI. Petrov had probably experienced painful confrontations at different interagency meetings, when he would be bitterly attacked by bureaucrats from the space program and space industry.

Based on such experiences, Petrov defined the Space Research Institute as "the boy to be beaten."

"But Georgy Ivanovich, at least from now on, you are free from such humiliation," I said soothingly.

"Oh yes, now I can do whatever I wish! Maybe now I should put an end to my space activities and spend my time developing the aerodynamic theory of insects' flight, or build a theory of the Tunguska meteorite explosion."[3]

Every story he told me was accompanied by an unimaginable num-

3. That famous meteorite which fell on Siberia in 1908 and still remains a mystery.

ber of anecdotes and jokes, which greatly helped me build a vivid overall picture of the new, hostile planet on which I had landed.

"You, poor boy, will be dependent on every—even the most absolutely insignificant—bureaucrat everywhere, beginning with the Central Committee and Gosplan. Those fellows will tell you what you have to do. They will draft and approve every bit of the planning. You know the results of that planning? Our poor economy lacks, for example, simple toilet paper. We all suffer from the shortage of it. And do you know why? Because of lousy planning. These fellows in Gosplan have been counting the number of heads![4] They clearly have no brains if they think that the number of heads is equal to the number of asses."

With great difficulty, I was able—occasionally—to squeeze in a brief remark of my own. At this particular moment, I thought I could not resist suggesting a scientific explanation: "Dear Georgy Ivanovich, I heard another interpretation for the shortage of toilet paper. The cultural requirements of the masses are growing faster than the industrial output of the socialist economy!"

At the institute's inception, Keldysh appointed a deputy director to Petrov, a protégé named Gennady Skuridin. Almost illiterate in everything related to science, Skuridin tried to influence the important decisions on space projects and on the internal life of the institute. Often he would be supported by Keldysh over the head of the director of IKI.[5]

Petrov started a real war to dismiss Skuridin as his deputy. He was supported by an absolute majority of the space science community, not only inside IKI, but in other academic institutes and at Moscow State University. I remember a friend of mine, a university classmate and a space theorist himself, telling me, "Skuridin is our space community's Lysenko."

It was not easy to fire Skuridin. He was decorated with the highest award for space achievement, the Lenin Prize. Such a title provided a lot of protection in the system, in addition to the indirect and direct support he always got from Keldysh.

Soon after I took over the institute, I discovered how indispensable Budker's wisdom had been with respect to appointing and carefully

4. Per capita planning.

5. Later I experienced the results of the dreadful selection of his entourage—a selection that was so typical for Keldysh.

selecting every individual member of your team. Unfortunately, in my new assignment that wisdom sounded like an ideal—an unattainable theory. I had to deal with almost 1,200 employees hastily recruited by my predecessor. Undeniably, many of them were bright scientists or talented engineers and technicians.

However, these crowds were attracted to IKI by the name of the institute and the general aura around the early space program. Many were newcomers—sheer adventurers without a serious background in science or engineering. Some of them had patrons in the high echelons of the establishment. This particular circumstance, as I soon discovered, often kept my hands tied. In a desperate mood I compared my role to that of a captain suddenly taking over a troubled and wrecked ship in the open sea.

I could not seriously count on much help outside the institute. Artsimovich had already died. Keldysh was ailing after complicated surgery. He was noticeably debilitated. Even when I did have access to him, my initiatives and suggestions were often blocked by his assistants and protégés. Additionally, as it had been in the case of the resigned Skuridin, they had infiltrated the very top of the Space Research Institute.

The first and most important task I had to undertake in the administrative area was to bring together the existing board of directors and to find out what I could possibly do with the individuals who had been imposed on me as my closest deputies and aides. In order of "implant," number one was Yuli Khodyrev, my deputy on engineering issues and practically the technical director of IKI. Strongly built, he was a fifty-year-old, somewhat balding blond. He clearly lacked verbal skills when he tried to promote the ideas of perestroika and even cultural revolution inside the institute. However, I sensed that he had the bright head of an inventor and engineer.

Khodyrev's main blemish, according to his opponents, was his absolute lack of administrative talent. Later I was to confirm it from my own experience in dealing with him. However, in the meantime, I was able to greatly benefit as his student in spacecraft engineering and logistics. His background had been in the space industry. Before joining IKI he was second in rank at the biggest mailbox installation for radio electronic subsystems in Soviet space. Through his tutelage, I quickly learned how to operate with "watts of on-board power," "kilos of payloads," and "kilobits per second" (of information flux from the ground to the spacecraft and vice versa).

The rapidly expanding IKI was a natural asylum for Khodyrev, and soon he became an absolutely indispensable favorite of Petrov. The wide range of Khodyrev's technical expertise fortunately complemented Petrov's theoretical background. After the long-awaited successful expulsion of Skuridin, Khodyrev became first deputy to the director.

The second man I inherited was Major General Georgy Narimanov, a contemporary of Budker. He was an expert in the fluid dynamics of rocket fuel tanks. An old-timer of the space program, he knew every important person in the establishment. Clearly, he was closest to the bosses in the Central Committee, in the space ministry, and even to Keldysh personally.

Closeness to the sacrosanct headquarters of space industry and rocketry, with its indispensable involvement in military uses, had given Narimanov a chance to acquire the rank of major general. This promotion did not represent anything extraordinary in the atmosphere of the fifties and sixties. Quite a few technical people, and even scientists, were given such military titles as symbols of appreciation by the Ministry of Defense and by the government. But it gave him a certain prestige that was helpful in certain circles.

I was not much interested in the transfusion of rocket fuel from one vessel to another; hence, Narimanov's main contribution to my reeducation was in the area of the internal politics of the Soviet space community and its headquarters. He was clearly a priceless walking collection of different facts and anecdotal stories. It was from Narimanov that I learned the science of survival, invaluable for a swimmer in an ocean full of military-industrial establishment sharks.

Though I admired his diplomatic genius, I learned that he himself had been fired from a very important previous position, as chairman of the Scientific Technical Council of the whole Soviet space and rocket industry. He clearly did not like to specify what the actual reason for his resignation had been. Instead, he limited himself only to vague remarks about the "great war between opposing clans." Apparently his departure had been the result of a final settlement or cease-fire. I could feel, only indirectly, the still unsettled smell of powder left by this war that had not yet been explained to me. My new friends in the space establishment gently warned me that I should never send Narimanov to any important meeting if the minister of the Ministry of General Machine Building,[6] Sergei Afanasiev, was going to be present.

Despite this, I accepted Narimanov as a valuable member of my team, and I tried to define his area of responsibility and competence as that of a clever apparatchik. The institute obviously needed such a person for complicated navigation in dangerous and turbulent waters.

Besides, Narimanov's uniform played an important role in our need to negotiate and deal with the military on an everyday basis. I learned rather soon that the Ministry of Defense played a crucial role in running the infrastructure that supported any space launch. All of the country's launching sites and space polygons had been placed, from the very early days of the Soviet space program, under the supervision of the military. Soldiers were considered the cheapest and, at the same time, the most disciplined labor force, especially in the construction of launch sites in uninhabited areas with unfriendly and even hostile climates. The military did not object to such assignments since a major part of the launches, including test firings, were carried out on their contracts.

During the first few years of my directorship in IKI, General Narimanov was an important contributor to rebuilding the institute and an especially valuable shield for the director and the scientific staff of the institute. He protected us from the external world of the military-industrial complex.

The third member of the board, Colonel Georgy Chernyshov, was not hired either by me or by Petrov. In the Space Research Institute he represented the interests of the KGB. Most of the academy's institutes did not have such "souvenirs" from the security branch. The "privilege" of getting such a deputy was due to the close association of IKI with the space industry, which was considered highly classified.

As it would be for any other director, it was not easy to communicate, much less issue instructions, to a deputy who received his salary from a completely different agency. At the same time, as director I was largely dependent on Colonel Chernyshov in several critical issues. First and foremost among them was in the area of staff recruitment. In every case, Chernyshov played the role of politcommissar, responsible for the selection and approval of the cadre.

6. The ministry for the space industry.

Chernyshov was short but athletic. His polite manners apparently had won him assignments to work with the intelligentsia throughout his career in the KGB. Before joining IKI, his previous duty had been to take care of the souls of Bolshoi theater stars. According to his brief comments, I understood that as a delegate from the KGB he had had to accompany ballerinas on their foreign voyages. I even thought that the choice of Chernyshov for his job at IKI complemented his previous activity, making him the guardian of the country's two best assets: the legs of our ballerinas and the brains of our spacemen.

If Chernyshov wasn't very intelligent, at least he was a man with enough common sense, and even generosity and respect, for big science. In practical terms, that meant I was able to find a compromise with him for most of my suggestions about new staff members. After giving him the name of my next choice for a position that had even the slightest degree of importance for the institute, he would take it and come back to me within a few days with an expression on his face resembling that of a cat after gnawing on the bones of its victim.

"So, what is the prognosis?" I would ask.

He would reply, "Well, I was unable to find any contradictions in the applicant's pedigree, but there are some nuances around Article Five." That meant "Jewish origin."

I would then insist: "What's wrong with that? Don't you think we should care above all else about getting the best brains here?"

My colonel would then assess the situation. "Unofficially, I completely agree with you; however, I had a hint from one of my bosses that the institute has already exceeded its unwritten quota of such employees."

A bit more gentle pressure and finally we would agree to one more exception in the established quotas.

I was amazed somewhat later to find out that the Space Research Institute had acquired special fame as a result of these successful negotiations. Behind the scenes, sharp tongues called IKI the stronghold of the Jewish-Tatar mafia. So what; if it helped to build better science, not to mention justice, I was ready to accept the punishment.

In one of my frank deliberations at a tea party, Chernyshov later explained his reluctance, and even resistance, to employ candidates with the troubled Article Five: "You know, Director, in jobs like mine they are counting the number of people who apply for emigration. Those of my colleagues who supervise different institutions or mailboxes that would exceed certain unavoidable quotas could have a hard time."

The stories about Article Five, at the peak of the Brezhnev stagnation era, gave rise to a whole series of anecdotes and jokes ridiculing the practice established by the system. This is one example:

"Do you have any problems in your curriculum vitae?"

"I don't think so, but I am not sure about my great grandfather. He was a gangster."

The chief of the cadre department of the institution would say, "Don't worry about that. Let's see what else might create difficulties."

The poor candidate finally would say, "I don't know whether I would be acceptable for your institute. I am a Jew."

With sudden coolness, the boss would respond, "We cannot possibly have anything against that. However, the story about your great grandfather makes you unacceptable for us."

The next, and often much more complicated, obstacle to overcome in my interactions with Colonel Chernyshov was to ensure permission for foreign trips for staff members of the institute. The increased scope of international cooperation and IKI's role as the window dressing for the Soviet space program created the understandable necessity of sending many people to international conferences and meetings, and on business trips. The authorities—the caretakers of ideological immunity—classified all their "patients" into one of two categories: "exitable" or "nonexitable."

Even among the cadre who got through all the clearances and were finally appointed to such institutions, there was a category of people who were considered nonexitable. There was no official explanation for the definition of such a label. I had quite a few very important staff members of the institute who were eventually identified as belonging to this category. For a more detailed analysis of different cases, I had to conduct very delicate dialogues with my deputy.

There were two principal obstacles for foreign trips, he later confessed. One, rather rare after many checking procedures at earlier stages, was considered a genetically inherited "disease"—that is, the applicant had some very distant relatives living abroad.

The other widespread type of rejection (I managed to cajole from Chernyshov) sounded like: "There are some 'entries' in the person's file who is under consideration for a foreign trip. No one can eradicate such entries. They're registered and given the status of official."

"What do you mean 'entries'? What kind?"

"You should understand that there are people, ill-wishers, who send bunches of letters and memos to us, calling into question the loyalty or faithfulness of the applicant."

"You mean there is no way to investigate and eliminate denunciations that come from anonymous bastards? Couldn't these ill-wishers have been driven by envy and jealousy?"

One of the greatest tragedies of Soviet citizens, until the very recent years of perestroika, was that there was no legal way to overrule such accusations. Moreover, they would surface only distantly on such occasions as during applications for an exit visa.

In my case, as the director of IKI I had to find ways to circumvent the obstacles, to beat the system. "You have to understand that the highest priority of IKI is to represent, decently, the achievements of the Soviet space program abroad," I would tell Chernyshov. "Everyone should cooperate with the institute in that, especially if you yourself understand how ridiculous these contradictions are."

Such conversations went on for a few days. "In principle, there is a way to overcome the entries. It would require the intervention of a higher-level official at my department in the Lubianka. Could you, Director, approach this man yourself and use the same arguments you just made in our conversation?"

Even if I "cured" only a modest number of "patients" of the affliction of being nonexitable, I felt proud.

Through these and other efforts, the Space Research Institute was, in practical terms, able to run a full-scale international cooperation program, long before the advent of perestroika.

In conclusion, with respect to my rescue operations for such people, I can share an observation: the brighter the person, the darker (and possibly dirtier) the entries were in his file. The system established by Stalin and Beria, whose ghosts outlived even the Khrushchev thaw, capitalized in a very wicked way on the old mentality of Russians: envy.

No institution could survive without a designated person responsible for managing household and bookkeeping tasks. In the Soviet Union, such a job requires the special talent to operate in a system without internal convertibility of money, and within a system characterized by a great deal of corruption that is not only officially recognized, but even silently sponsored.

The "deficits," the famous shortages of almost every vital item, required special administrative skill: to make deals without *too* much violation of the rules. I can swear, based on my long experience as director, that it is much more difficult to find the appropriate candidate for this role than to hire a dozen bright scientists.

My very first day in the office, I was introduced to the deputy director responsible for household management, Anatoly Ivanov. I was told that IKI had searched everywhere. "Nobody in Moscow was able to fit our requirements," I was told upon my arrival. "Finally, we found *him*, four thousand miles from here."

Although he had only been in the institute a few months, Ivanov was in control of everything. Full of energy, he started several innovative projects to strengthen the material and economic base of the institute.

Ivanov was an expert in the art of making deals. At the same time, he was absolutely honest and untouched by corruption—not a small consideration for me at this particular moment. As director, I had to worry about the impression the institute would create on the soon-to-be-expected crowd of American spacemen, engineers, and—as I still hoped at that time—scientists, who were coming as part of the forthcoming joint flight and docking between American and Soviet spacecraft: the *Apollo-Soyuz* project.

Great effort and nervous energy were required of me at the beginning of my director's career to prepare for another visit of American guests, as an important goodwill gesture of the détente era. This diplomatic miracle was the result of détente, coupled with the intention of the two superpowers to cooperate in space. Essentially, here's what happened:

An American high-altitude balloon for cosmic ray physics had been grounded by Soviet Air Defense. The parachute and gondola were eventually recovered by the military, and the American Embassy in Moscow was informed that the Soviet side was ready to hand over "every bit of what is in the hands of Soviet authorities."

My institute was given the enormous honor of delivering this "gift." It was a rare attraction—a curiosity—and all of us in the institute immediately went to take a look at what we were going to present the next day to the American officials. It was a huge, bright parachute of orange color, with a gondola about one meter long. The structure carried a rather unimpressive payload, at least for a layperson. However, the political significance of the coming event was enough to attract a dozen newspaper journalists and reporters for the next day's ceremony. And all of us were looking forward to the event.

Early that morning, however, I had a surprise call from Colonel Chernyshov.

"Director, I am very sorry, but we are in big trouble. Someone during the night committed an international crime. There is not a single trace of the parachute silk." It had disappeared together with a piece of hardware later identified as a photographic camera for cosmic ray experiments.

Half an hour later, at an emergency meeting of the institute's board, we discussed different hypotheses on what had happened. The amateur detectives among us immediately suggested that the thieves, in any case, were not professional scientists or technicians. Only a completely illiterate person with no knowledge of optics would have been attracted to such a camera, whose use was limited only to a particularly narrow cosmic ray experiment. The huge piece of specially strengthened parachute silk, of course, was another matter. We knew its desirability, and we all felt a real sense of humiliation that one of our countrymen could have been so greedy. At the very beginning of a bright international career for the institute, the incident was a knife in our back: it had the potential of becoming an international political scandal.

Chernyshov said, "Alas, my only responsibility is to keep Soviet secrets."

The housekeeper Ivanov was clearly suffering from the loss. Even his body language showed how offended he was, and he promised us: "I will check with every employee of my division. We will be able to reconstruct the developments of every hour of last night."

After all these years, I am still unable to reconstruct what happened during the very next hour in the institute. However, I remember very well that while I sat in my office in a dark mood, I had to accept phone calls with unpleasant comments from almost every corner of the official establishment: from the Academy of Sciences, the Council of Ministers—even from the Central Committee.

Suddenly, I saw Chernyshov enter the room with the expression of a contented cat.

"Director, stop worrying. Everything is clarified now. The mystery has been resolved. We have recovered the silk."

I don't know how much Chernyshov's professional skills had been involved in making the final deduction, but the truth of the revelation really shocked me. My deputy, the housekeeper Ivanov—brought to Moscow as a rare combination of managerial skills and personal virtue—was the person who had taken the parachute. As he explained it to me, he was acting according to simple practical instincts: the Americans would never use the same parachute again, and Ivanov had had an urgent need to protect his personal car with a cover. The parachute silk—the souvenir from the sky—was perfect for that purpose.

Ivanov was absolutely the first man I fired in my life. I never heard from him again. However, somewhat later Chernyshov told me that Ivanov still retained his security clearance, important for his next employer.

The IKI was chosen by the government as the "open site" for the *Apollo-Soyuz* project. We were the hosts of the working meetings and negotiations that eventually led to the final tests of the docking modules. The huge Soviet space–military-industrial "iceberg" at that time could find only a tiny place on top of this colossus to deal with foreign visitors. Even if IKI was unable from a practical standpoint to serve as the sole counterpart to the whole *Apollo* team of NASA, we were given instructions to at least pretend. We had to puff up our chests and represent the heart of the Soviet space program while it was very clear, even for amateurs, that we were nothing but a bunch of scientists—the poor relatives of the rich space czars.

The principal negotiators—technical experts on space technology from the Soviet side—were introduced to our American counterparts as staff members of IKI, though some of them had only an hour before changed from their military uniforms to civilian clothes. I don't think our guests were simpletons who accepted everything at face value.

I remember the visit of James Fletcher, then chief administrator of NASA (National Aeronautics and Space Administration). We had a very nice and polite conversation. He probably knew everything, but I felt that he didn't want to embarrass me by asking tough questions about space technology and the operations involved in the upcoming project. Instead, we had a nice conversation on how to run in-house restaurants or cafés for employees, and how to organize maintenance on the elevators or centralized electric clocks on the wall. It was a great relief for me to leave aside the complicated technical questions that lay beyond IKI's expertise. On the other hand, it was nice to feel that my institution was not alone, trying almost desperately to survive against the logistical issues associated with running an institution.

Well before the critical moment of the *Apollo-Soyuz* project, when the crowds of American participants went to visit the control center for manned flights outside of Moscow in Kaliningrad and *Soyuz* was launched at the Baikonur Cosmodrome, our authorities issued a secret document—a very long and detailed questionnaire.

In a form accessible even for a stupid person, it prescribed everyone's responsibilities. It also suggested the exemplary answers to hundreds of questions that might be asked by those "nosy Americans."

The reading of this questionnaire at the board of directors brought great joy and pride to us. These were some of the questions:

"The imaginary American asks, at the control center of Kaliningrad, 'Who is essentially running this installation?'"

Or a similar question at the Baikonur Cosmodrome, where military servicemen would have most probably just changed their uniforms for civilian clothes: "Who is responsible for supervising and running this Cosmodrome?"

We were very excited: the recommended answer was always "IKI and academician Sagdeev." I felt like the Marquis de Carabas from Charles Perrault's fairy tale, "Puss in Boots"—the apparent owner of all the territory the eye could see.

Fortunately, our sense of humor was developed enough to think this document genuinely funny. I believe our American counterparts on the project, even without being able to read that entertaining manuscript, also enjoyed and laughed at the answers of their courteous Soviet hosts. At the same time, I think that on the Soviet side the real participants of the project from different space installations felt uncomfortable at practically every meeting with the Americans. They had to physically come to IKI for everything, even if they had to make intercontinental phone calls, which was necessary several times a week.

Poor Konstantin Bushuyev was chosen as a technical director of the *Apollo-Soyuz* project from the Soviet side, making him the man who took most of the trouble for himself. Since he was officially introduced as an employee of IKI, he was obliged to follow this fantasy even after the project was over, or any time we had an unrelated follow-up visit from NASA. Moreover, even Bushuyev's body became part of a post-mortem masquerade. After his death, instead of having a decent funeral at the former Korolev Design Bureau, where he had spent most of his active working life, the final sad ceremony was moved to IKI, simply as a cover for the "Marquis de Carabas" scenario that had been staged so many years before.

"Suppose these nosy Americans try to attend his funeral," the military had reasoned.

To us scientists, this show was a real sacrilege. The *Apollo-Soyuz* mission had nothing to do with science. Contemplated as a part of the détente package in 1972, the project was a quick demonstration—a handshake in space. It was something that could only be of interest to politicians. The engineers who actually carried out the project tried in their turn to simplify the actual scenario of the docking and the joint flight.

I believe that Bushuyev himself felt uneasy about denying science the opportunity to create a few serious scientific experiments on this flight. He used to tell me, "Please wait; be patient. The very next joint project will definitely be one of substance."

Unfortunately, Bushuyev didn't live long enough. I, on the other hand, had the really good fortune to be part of many joint projects, even if they did not come to us as a result of top-level summits. After all, the general political climate deteriorated after a brief détente, and in many ways everything we space scientists were able to get came independently of the government. Sometimes we got what we needed despite the big bosses and we took the risk of pushing for cooperation, while official circles were involved in the rhetoric of the Cold War.

22 / The Chief Designer and The Chief Theorist of Cosmonautics

➢

Not long after I assumed the position of director of IKI, the president of the Academy of Sciences, Mstislav Keldysh, underwent complicated surgery. He never recovered as an active administrator or scientist.

The old-timers in the scientific community had told me many stories about his important role in the early space program. Canonized in the official legends, three people were made icons: Sergei Korolev, whom the authorities acknowledged as the chief designer of cosmonautics only after he died;[1] Keldysh, the chief theorist of cosmonautics, who was lucky enough to live somewhat longer and was therefore able to be officially identified as bearer of that mysterious title; and finally, Yuri Gagarin, the first man in space.

During my time at IKI, Keldysh kept all of his official regalia, even though he was unable to meet the real obligations of a leader of the space program. Formally, as president of the Academy of Sciences—the headquarters of Soviet science—he had important say over anything related to high technology.

I was constantly in touch with Keldysh, from the time he appointed me director of the institute, until he died five years later.

There were periods during those brief years of my interaction with Keldysh when he called me almost every day, inviting me to visit him. The typical format of such meetings was informal. There were very few of us who were asked to spend hours and hours in Keldysh's office, drinking tons of coffee. After his resignation from the presidential chair in the academy, he invited me to the office where he served as director of the Institute of Applied Mathematics, a position he kept until his death in 1978.

1. Korolev's place of honor was to be succeeded many years later by Valentin Glushko.

As intellectual leader of the space program, Keldysh was also the principle user and key holder of the best Soviet computer technology. This last duty was associated with the role of his Institute of Applied Mathematics in both the Soviet nuclear and space programs. He tried very hard and did not spare himself. In fact, he burned his heart and body in the high fever of his ambition. Nominally, he was decorated with every imaginable and unimaginable award—medals and prizes that only the Soviets could issue—and he was honored with numerous foreign awards. But deep inside his soul, I believe, he felt he had been a failure.

When Keldysh became prominent in the space program, which had flourished in the late fifties when he was second only to Korolev, the atmosphere of nonstop victory suddenly changed to a painful recognition that the Soviets were losing the space race to the Americans. During one of our very open and frank tête-à-tête conversations, I said to him, "If you really feel that we are so far behind, why don't you talk to Brezhnev? Something has to be done."

With a sad smile, he answered, "You think Brezhnev doesn't know it?"

If the good Lord had the special intention to test and temper me during the worst moments of Soviet space history, there would have been no better time for that than in the spring of 1973. In April of that year, only a month after I had assumed office in IKI, a huge module for the planned orbital space station *Salut* almost exploded in orbit over the Pacific Ocean. After only three weeks in space, without ever having accommodated a crew on board, an expensive toy was lost.

The collapse of a twenty-ton piece of hardware could not go unnoticed, and the leaders of the space establishment had to provide an explanation. The wise men of the space industry suggested as a most plausible excuse an encounter with a meteor. Those who were not so gullible understood that the probability of being hit by a meteor is one in a million. An accident like this would have required all the evil forces of nature united to create such an improbable occurrence.

In my new office for only a few weeks, and still recovering from the shock waves left by incidents like the one of the American high-altitude balloon, I had to report an even worse discovery. Four unmanned interplanetary probes, called *Mars 4*, *-5*, *-6*, and *-7* respectively, which were under last-minute preparations for launch in July of that year, were discovered to be doomed.

Unnoticed at the time, tiny violations of technological procedure in sealing the special space-qualified series of microchips had left all four spacecraft with hundreds of defective chips. A technical flaw of this nature would create the unstoppable process of corrosion, which would destroy some of the microchips in flight. It was like having a time bomb or an incurable virus. The only question was how long "the patient" could survive.

On such short notice, there was no practical or theoretical chance of replacing the hardware on board the spacecraft. The best brains of the space establishment were mobilized to predict the possibility of the spacecraft's surviving a six-month trip to the planet Mars. Unfortunately, the forecast was pessimistic: fifty-fifty. Despite these odds, no one could reverse the momentum of the countdown to the final launch. In that sense, the conscious decision to fire was equivalent to playing a space version of Russian roulette.

What happened a few months later constituted the single biggest disaster in the Soviet space program. Two of the four planetary probes lost their ability to maneuver on command from the ground. The result was that they simply missed Mars. The two remaining spacecraft were largely debilitated. One of them, programmed to make a soft landing on the surface of Mars,[2] did not survive touchdown. Its otherwise invaluable data on the Martian atmosphere, which had been accumulated during a direct-descent trajectory, reached the ground station in unreadable form. The virus was everywhere. Almost every piece of hardware and scientific instrumentation on board malfunctioned.

IKI, as principal scientific customer of the project and now the unlucky recipient of a miserable amount of scientific gibberish, was in a state of mourning. However, that was probably not enough to punish its still-innocent director. I was asked by the authorities to give the press conference at the conclusion of the flight. I only remember how embarrassed I was by the painful cross-examination conducted by prying journalists. The confidentiality of the problem made it impossible to tell them about the microchip virus. Clearly such a disaster was incompatible with the "healthiest space program in the world."

In a state of utmost desperation, I interrupted the intrusive inter-

2. This probe was a precursor to the American *Viking* landing.

viewers: "Look. Don't ask me any more questions. Yes, I myself know the truth, and if it depended solely on me, I would tell you. However, I have to follow the rules that were established by our space community long before I joined it."

Apparently such an answer amused the foreign journalists. Some of them commented later that it was the first time they had encountered such frankness in talking about Soviet space launches. That was small consolation, since the avalanche of real mishaps was continuing to develop.

Keldysh invited me to his office the next day, and with a sad, maybe even ironic smile, suggested that I join a small gathering, symbolizing the funeral of the N-1 project. Facing my blank perplexity, he said, "We have to cancel all the programs and contracts that have remained since the failure of the lunar superlauncher."

The N-1 project was an old program—from the time of Korolev, whose mission had been to send a Soviet man to the moon before *Apollo* could deliver the American astronauts. Korolev himself died before critical tests of the booster could take place. Mishin, who succeeded him, failed three times in attempting to launch the rocket. One of the explosions literally smashed the expensive installations at the launch site.[3]

After Keldysh briefly introduced the regrettable story of the unfulfilled mission to the moon, my first reaction was: "Why did you wait so long? The Americans are already back after several trips to the moon, and have almost forgotten about it. Why didn't we close the N-1 project long before?"

I was very grateful to Keldysh for his frankness, and for explaining to me the heart of the problem, even when it required beating me over the head.

"Young man, you should understand simple wisdom. Nobody wanted to cancel it. It would have meant confessing before the government the failure of this expensive project, probably worth two billion rubles. The documents we have to sign now are very crafty. You will find out how smart we all are. We are closing the N-1 project not as losers, but as ingenious people who have found a much cheaper and more intelligent way to explore the moon."

3. A few months later, on my first trip to the Baikonur Cosmodrome, I saw the deformed and burned remnants from a distance.

"Mstislav Vsevolodovich, do you mean the successful series of small unmanned probes sent to the moon to return with lunar samples?" I asked.

"Yes, of course," answered Keldysh.

By that time, I was no longer a simpleton who would buy such an explanation. I immediately interjected, "But these are two completely different genetic lines in the program. The people who design and send unmanned rockets are indeed the heroes, except they have nothing in common with the N-1 team."

Keldysh looked at me seriously and said, "There is an opinion at the top that it would be better for all sides, including the future space program, not to create negative emotions."

I was completely stunned by such official hypocrisy. These people were capable of surviving only in an atmosphere of secrecy. Later I thought that maybe the wise men had actually been the ones who introduced the curtain of secrecy themselves. That way they could get out, unscathed, in the event of failure. If that were the case, I thought, this system which cheats itself is doomed.

On the technical side of the N-1 project, the final discussion convinced me that there were two principal deficiencies in the design and in the scenario of implementation. In order to launch it into an intermediate orbit around Earth, a payload heavier than a hundred tons would have been needed, as a minimum, to start the lunar expedition. The first stage of N-1 was built on the assemblage of fifteen or sixteen individual rocket engines.

Almost everyone in the space industry, as I learned later, was doubtful whether it was possible in a short time to beat the Americans. Moreover, the leaders of the project were in such a hurry that they did not dare schedule a comprehensive program of tests, which would substantially reduce, if not eliminate, the risk of blowing up the huge and expensive construction at the launching site.

Despite the failure of one of the most expensive technical projects in Soviet history, Vasily Mishin, who by that time had been promoted to the rank of general designer, still felt that he was in an invulnerable position. The very formula of the final funeral document for N-1 was designed to preserve Mishin's role. In the meantime, Mishin, who controlled the Soviet manned space program, was quickly building the capability to keep in orbit several cosmonauts at once—on board the space station *Salut*. Mishin was promoted.

The opportunity to represent the Soviet Union in the *Apollo-Soyuz* joint project, of course, had also automatically fallen in the hands of Mishin's design bureau. I visited Mishin's headquarters several times for huge gatherings that were convened to define the future scenario and programs of manned flights. Unfortunately, I wasn't given any reason to change my earlier Novosibirsk assessment of Mishin for the better.

As ambitious and expensive as Mishin's program was, it left almost no room for serious science. I felt like a freshman recruit taken to the front line, trying to promote the interests of my institute under hopeless circumstances. Even Keldysh, an experienced "space wolf" and nominally above Mishin in rank, was pessimistic about the real benefits of Mishin's program to the Academy of Sciences.

Mishin's mysterious buoyancy was explained thus. Despite obvious miscalculations and wrongdoing, Mishin had hidden support from an important protector in the Politburo, Andrei Kirilenko. He was the second man in the party ranks, whose son-in-law, Yuri Semenov, quickly appeared on the space horizon as the rising new star on Mishin's team.

At that time, I hadn't yet had the chance to assess whether there was real merit behind Semenov's promotion. However, I made some sad philosophical conclusions. Suppose two competing directors or designers hired, independently, the kids of top bosses. The way the Soviet system operated and created privileges, the boss would be immediately involved in the competition. Who would derive the benefits of employing the princes? Clearly, the one who promoted his princely protégé faster. Such an algorithm, however, would create unlimited escalation. The offspring of party bosses would take over the most important positions, sooner or later.

Milovan Djilas, who in the early seventies was close to Yugoslav president Marshal Tito, became known as a dissident after his book *The New Class* was published. Djilas's theory described how the ruling class perpetuates its domination in an atmosphere where the errors, even the serious ones, are not recognized. Crime, therefore, is not punished.

Although Mishin seemed to be a cat with nine lives, there was really nothing that could have foretold his quick fall. Only a few people knew he was a marked man. When Keldysh hinted that to me, he added that the explanation was much simpler, and it had nothing to do with the failure of the N-1 project.

The ill-fated superbooster had exploded, wasting thousands of tons of rocket fuel, but that had not harmed Mishin at all. But the poor fellow fell victim to smaller quantities of "fuel," better known as cognac and vodka—the stuff he used to boost himself into his own self-propelled orbits. (His deadly habit, which manifested itself long before Mishin's visit to Novosibirsk, would eventually kill him.) With it, he compromised his patron and protector in the Politburo. Even the letter sent to the Central Committee, signed by a group of faithful and subordinate sycophants in his design bureau, including Semenov, didn't help much. The search was on for a successor to that key job. Out of the whole Soviet space empire, it was equivalent to the nomination of heir to Korolev's crown as rocket czar.

Finding Mishin's successor was the most important event in 1974. The procedure was kept secret behind the closed doors of the government and party establishment. At the beginning, when I heard the name of the competition's lucky winner, academician Valentin Glushko, I was rather unmoved. While I had never had the chance to meet this academician, I vaguely remembered his face in the gallery of "dead souls" that was the pride of Sheindlin, the director of the Institute of High Temperatures.[4] Judging from Keldysh's reserved reaction, I understood that my boss was also not enthusiastic about the choice. But he probably had substantial reasons for his attitude.

My "know-everything" walking encyclopedia of Soviet space, General Narimanov, was much more forthcoming. He was absolutely knocked out by the news about Glushko.

"Look," he said to me, "Valentin Petrovich clearly is an extraordinary man; he is a man of one love, and one love only. From the days of early childhood, after reading the popular books of Konstantin Tsiolkovsky, he became a converted fanatic of space and rocketry. His entire life has been dedicated to inventing and designing rocket engines that operate on liquid propellants. In the narrow field of rocketry, he is the greatest expert in the world. However, his nomination as heir to Ko-

4. I thought how curious it was that we were at that moment hanging together on the wall of the scientific council's room in the Institute of High Temperatures.

rolev, even eight years after that man's death, is a sacrilege. Everyone in the country knows that the two men never liked each other. They were rivals, even enemies. Did you hear the *sharaga* story?"

I said to Narimanov that the name Glushko, as well as the name Korolev, had been cited by Rumer, my dear friend from Akademgorodok, who himself was a survivor of the *sharaga* and the gulag. Both men had politically supported Rumer many years later.

"Way back in the years of the *sharaga*," Narimanov continued, "Glushko was, according to legend, slightly more superior than Korolev. Or, to put a different way, a slightly more privileged intellectual serf. Then, for many years in free postprison life, Korolev was number one. He, not Glushko, was the chief designer of cosmonautics. Few people in the country had even heard of Glushko—imagine how much this hurt his feelings! He developed a strong inferiority complex and a mortal jealousy. Is there ultimate justice or not? It was he, Glushko, who was in correspondence at the age of fifteen with the grandfather of cosmonautics, Konstantin Tsiolkovsky. It was he who had the blessing of the founder of rocketry, not Korolev."

Narimanov continued with conspiratorial delight: "I can make an important prediction. Glushko, having acquired such a top position, is going to rewrite the whole history of the Soviet space program."

I cannot recall whether I accepted such a prophecy at that time, but many years later, when General Narimanov himself had gone, I was amused to get a personal gift, a book sent to me by Glushko, which literally confirmed Narimanov's clairvoyance. The red leather cover of *the Encyclopedia of the Cosmos* introduced Glushko as an unprecedented figure in the history of space and rocketry, citing how everything in his bright career had started with an omen and then Tsiolkovsky's blessing.

I was touched to get such a souvenir from the editor in chief himself, academician Valentin Glushko.

Since the moment of Glushko's ascension to the throne, all of us—Keldysh, myself, and our colleagues— waited with great impatience for a new initiative to cure the troubled Soviet space program. However, the scientific community, as an insignificant distant relative of rocket technology, was not part of the early brainstorming led by Glushko on his new estate.

In some sense, his actual power was even greater than Korolev could ever have dreamed. He was in control of both the former Korolev estate and his own former company, the stronghold of rocket engines.

Keldysh invited me to attend Glushko's first presentation, prepared for a narrow circle of people in the office of the president of the Academy of Sciences. The spectacular color posters told us everything, even before Glushko had a chance to open his mouth. We witnessed his almost science fiction vision of a lunar expedition delivered by heavy rockets to a base on the surface of the moon. There cosmonauts would have a chance to spend an extended period of time and perform different experiments and tests. To justify such an expensive undertaking, this particular new component of the suggested scenario had been designed as the Soviet response to the *Apollo* landing.

The centerpiece of the proposal put forth by Glushko was the newest heavy booster, bigger than *Saturn 5*—the workhorse of the American *Apollo* missions. However, in an odd way I felt that the main goal of the vindictive Glushko was to beat the ill-planned disastrous N-1 booster initiated by Korolev. Glushko simply couldn't stop competing and fighting with his old rival. And now he was looking to us, his stunned and impressed audience, to support his vision and to complement it with the necessary scientific justification.

The debate began, and with it many dozens of hours were spent and hundreds of cups of coffee consumed. That much we had to sacrifice with Keldysh to develop indisputable counterarguments. In the beginning, I was irritated with Keldysh's insistence that we go into every detail of the superproject.

"Mstislav Vsevolodovich, why should we waste so much of our time?" I asked. "It is clear, even for a child, that this is a ridiculous, premature project. How can we involve our country in such an expensive undertaking? After all, didn't we use our propaganda to convince everyone that the Soviet 'special way' of lunar exploration, based on unmanned and cheap robotic missions, is much better?"

I believe now that Keldysh had been smart enough to take his time. He was in no hurry to support Glushko without additional arguments from our improvised commission. In conclusion we finally said no to the project and gave quite substantiated reasoning. Fortunately, Keldysh's influence and the value of his judgment were still highly regarded by the government. Our recommendations were accepted.

In a desperate kamikaze attempt, Glushko produced a collective letter signed by six astronomers and geophysicists—members of the academy—in which they expressed their discontent with our assessment and highly praised the renewed lunar expedition. Knowing at least some of the cosignatories, I was sure that they had no personal interest in such a program. They had simply fallen victim to manipulation.

Almost ten years after Keldysh died, in the mid-eighties, Glushko tried to revive the project once more, after the very first publication of the American idea of a permanent lunar base. I was not surprised by Glushko's renewed attempt to raise this issue. However, I was rather surprised to get an invitation from Colonel-General Aleksandr Maksimov, head of main space directorate of the Ministry of Defense, to attend Glushko's presentation of the idea. It would be even more ridiculous if this time the military served as the customers of the project.

However, this time it did not require dozens of hours and innumerable cups of coffee to reject Glushko's proposal. It was a very brief meeting, one or two hours, and that was all. Glushko had lost most of his eloquence.

Glushko had an extraordinary career in the space program. The blessing by Tsiolkovsky launched him into a record term of active space life that lasted longer than sixty years. Imagine the thirteen-year-old boy passionately making a tiny toy rocket, and sixty-seven years later launching *Energiya*, the heaviest booster on Earth, driven by almost 2,000 tons of rocket fuel.

23 / The Military-Industrial Complex

➢

Did we have a military-industrial complex? We knew the Americans had one, because we first heard of it from President Eisenhower in his farewell address in 1960. The inner logic of it seemed very clear to us. As the Monopoly game defines a market economy, we could immediately imagine what kind of games were played behind the closed doors of the American complex.

We learned even more from numerous interpretations given to us by our own homemade propagandists. Nevertheless, there was absolutely no doubt: nothing like it could exist in our own society. By definition, our defense system was organized on different principles. After all, for many decades we had even been unable to find out if we had a military budget at all.

The early story of the Soviet defense establishment couldn't be separated from the political environment of the country, which considered itself an island of socialists surrounded by an ocean of hostile capitalists. It led to the implementation of a specific kind of psychology, based on survival against external aggression.

Every day, every month, years of indoctrination finally instilled a siege mentality in the people, who eventually accepted the need to build a strong defense. From the very beginning, the leaders of the country understood that a strong defense would require development of military technology. In the twenties and thirties, this task was in the hands of the young military generals and leaders who took care to build their military industry despite the country's struggle to provide the bare necessities for its people. That generation of military leaders was rather farsighted. Marshal Tukhachevsky himself had supervised not only military aircraft designs, but even the early experiments in rocketry—even though it was not then obvious that they would be so indispensable for military purposes.

Stalin, in his internal political struggle, established an unchallenged position as dictator and eliminated all these bright generals. Like Tukhachevsky, many of them perished just before the war. In many respects it explains to me the disastrous beginning of World War II against the

Nazis. Not only was the leadership of the army unprepared and left without its best minds, but the development of the military industry was damaged and paralyzed. However painful and bloody it was, the war did eventually help the military-industrial establishment recover.

During the war everyone in the country acted according to the slogan—I still remember it very well—"All for the front, all for victory." With such a slogan, attempts to rebuild the military and industrial sectors were given an understandable priority. The defense companies, having hastily evacuated from the European region of Russia to the Urals, were considered the most important enterprises. Their workers were given special food rations that exceeded the norm in the country.

The preeminence of the military's industries and its science was strengthened with the advent of the nuclear age. Trying to catch up with the Manhattan project, the government established a special committee led by Beria to supervise such projects.

With the fall of Beria, the defense establishment had quite a few notable supervisors, all of whom represented the highest echelon of power, the Politburo. For a number of years under Khrushchev, Brezhnev himself was a commissioner for high technology or rocketry.

Sakharov was still at his hydrogen bomb installation when he was summoned several times by Brezhnev. Brezhnev understood, of course, the importance of rockets, not only as something for the military, but also as a very important political weapon. Khrushchev continuously demonstrated his interest, using every possible photo opportunity to be seen with cosmonauts like Gagarin or, later, Gherman Titov—his "space brother," as people joked. In my space career, when I had to deal extensively with the defense industries, rockets in space were provided as a show of philanthropy from the military's enterprises.

Dmitri Ustinov, secretary of the Central Committee and the member of the Politburo responsible for the defense industry, was a technocrat. Graduated as an engineer, he had worked for years in the tank industry. Ustinov's talent as a manager and his ability to quickly learn how to work under the strongest pressure were noticed in Moscow. He was promoted to higher positions, and when the government rearranged its network of defense industries, it created a special body that was to be part of the Council of Ministers. Called the Commission on Military-Industrial Issues,[1] it was, by charter, led by the first deputy to the prime minister.

1. With a name like that, I was never able to understand how our propagandists could ever have denied the existence of our own military-industrial complex.

Ustinov started this new organization. With the growing scope of military research, development, and production—including procurement—it became a huge bureaucratic agency and he accrued substantial political power. This power finally propelled Ustinov to the top of the Central Committee of the CPSU: its Politburo. In this position he did essentially the same job he had done for the Council of Ministers—controlling on a day-to-day basis hundreds of companies and enterprises belonging to different ministries. His supervisory powers extended, however, to the political stronghold of the nation. His previous chair was given to one of his former siblings in the military industry, Comrade Leonid Smirnov.

Outside the military-industrial empire and party circles, very little was known about either Ustinov or Smirnov. They were gray cardinals. Together with my own acquaintance from the Geneva conference, Ivan Serbin, who had kept his position as head of the Defense Department of the Central Committee, they formed an influential trio. Nominally, Serbin had to report to Ustinov, but in the complicated internal political fabric of the Central Committee, he could be summoned directly by General Secretary Brezhnev.

Before I took my position as director of IKI, thereby joining the space elite of the country, I had no knowledge about this enormously influential hidden part of the military-industrial iceberg. It was hinterland to me. I don't think Artsimovich and Budker were regular clients of this privileged society in the country; their interaction with that iceberg was most likely limited to the Ministry of Medium Machine Building, which nominally ran Kurchatov Institute, a part of the nuclear energy establishment. Since Kurchatov had moved direct military applications from his headquarters to other "installations" scattered throughout the country, like Arzamas-16 or Chelyabinsk-70, the leading members of his team were probably liberated from day-to-day contact with the then-hidden parts of the military-industrial establishment.

From my very first day at IKI, I had the feeling that the power base—the actual leverage over the whole space program, including the tiny part represented by the space science community in the Academy of Sciences—was critically dependent on some undiscovered mysterious headquarters or agency. The most important secret documents were kept in special storage with other classified materials in the so-called First

Department of the Commission on Military-Industrial Issues. From time to time they were sent to the Space Research Institute.

Though all of this secrecy and confidentiality sounded tremendously significant, the content of the papers sent to my institute had nothing to do with any kind of military secrets, weapons, military hardware, or munitions. As a matter of fact, the first important messages of this type to the institute were related to the soon-to-be-launched unmanned Martian space probes, *Mars 4*, *-5*, *-6*, and *-7*. There was nothing classified or considered secret in that mission except the disastrous story of the doomed microchips. Every bit of the scientific payload, the ultimate messenger to Mars, was built by my institute or its likes. Every bit of instrumentation was assigned to perform normal scientific experiments, like the imaging of Mars, taking samples of gases in the atmosphere for analysis of the chemical composition, or experimenting with plasma physics in the planet's distant environment.[2]

Moreover, everyone in my constituency knew perfectly well that the very day after the launch of this cavalcade to Mars, all the newspapers, radios, and TV stations would immediately report the event. Anything scientific that was brought back would eventually be published in open scientific literature. The whole secrecy around such a purely scientific mission reminded me of the children's game I've Got a Secret. Only later did I understand the wisdom behind such secrecy. In the meantime, I took it as an almost natural consequence of the very fact that a major part of the hardware, beginning with the launchers and the spacecraft's systems, had to be produced by the Ministry of General Machine Building, the home of Soviet rocketry.

Very soon I started to get phone calls and even visitors from that mysterious Commission on Military-Industrial Issues. Introducing themselves, these men never cited the long and almost unintelligible name of their organization. Instead they introduced themselves as persons from VPK, the literal translation of which would be MIC, military-industrial commission.

I was struck by the symbolism. It could be the acronym for the military-industrial complex, as well as the commission. The way my

2. The atmosphere of Mars borders on continuous streams of plasma of solar origin, known as solar wind.

new friends used the acronym indicated that they had nothing against being openly considered representatives of the respectable military-industrial complex. However, the very fact that the acronym was held in such secrecy had a very simple explanation. The government, even the ideologues, admitted quietly among themselves that our country also had something similar to the military-industrial complex as defined by President Eisenhower.

In the summer of 1973, a few weeks after my appointment as director of IKI, I was summoned in person to the Kremlin to visit this mysterious establishment known as VPK (MIC). Following verbal instructions received by phone, I reached the famous Spassky Tower, whose huge clocks and bells signified for many years the beating of the country's pulse under the Kremlin's Communist leadership. Very close to the place I was to go is St. Basil's, the ancient monument to Ivan the Terrible's conquest of my native Kazan.

I had to enter a small one-story building, almost a booth, to produce my academy ID. The clerks behind the windows carefully studied my documents, comparing them with some files they had. Before I was given an open page pass, the voice of a military serviceman—the clerk—explained to me what to do next.

At the gate, I was to produce my pass and ID for a few seconds of quick assessment by the officer on duty. Then I would enter the Kremlin's interior. After turning to the right, I had to walk a few hundred feet, passing a long, dull building before reaching the place described as my final destination.

It was an old palace with a rotund entrance and a staircase elegantly bent in the form of a helix. While the officer carefully studied my documents, I thought: "What symbolism! The medieval stronghold of the Russian czars and autocracy is converted now into the fortress of the Soviet military-industrial complex."

In the reception room I found a large crowd of busybodies. There was an atmosphere of impatient expectation—almost alarm. I realized that a big imposing door leading to another room was the center of attention. There was traffic in and out of that door every ten or fifteen minutes. Those who were waiting to go in were impatiently asking those who had just come out, "How did it go? Were you laid on your ass?"

Some replied with brief remarks like: "Don't even ask—it was a rather severe scolding."

Or in rare cases: "No, I was born in a lucky shirt."

The secretary would then loudly announce the topic of the next meeting behind the doors, and the next group of potential victims would swallow a large gulp of air before entering the room.

My new friends, the clerks from the commission, like basketball coaches before sending their boys to the game, gave me last-minute instructions: "Stay calm. Don't ask new questions. Don't let them get excited or suspicious. Don't let them focus their attention on sensitive, subtle items."

I entered with a group of chief designers and officials from the space industry enterprises who were participating in the forthcoming Mars expedition. We entered a huge room that had columns at one end and the traditional white marble Lenin sculpture at the other. In the center of the room, there was a large horseshoe-shaped table occupied by, as I quickly sensed, the most important figures of the military-industrial establishment. On the periphery of the room were several rows of chairs for permanent aides and staff members, and another group of chairs for visitors introduced for specific items only. Later I was told that the man who occupied the top chair was Smirnov, the head of the commission, and his companions, mostly ministers of industries classified as the "indispensable part of the defense establishment." We called them the "Magnificent Nine."

I was not able to name all of them from memory. I knew, of course, a few of them with whom I had to deal, beginning with the Ministry of Medium Machine Building, my old acquaintance from the years at the Kurchatov Institute. There was also our current "big brother" who encompassed all the activities for the space industry: the Ministry of General Machine Building. Then there was the Ministry of Radio Industry, working in the areas of electronics, military radar, and computers.

There was one that had the straightforward name Ministry of Defense Industry. It was said to be the oldest ministry, and so it had the right of seniority to retain a privileged title. In fact, what they mostly did was produce tanks. Curiously enough, however, that very ministry was the principal contractor for all the optical equipment in the country. So, as a space institute we were dependent on them whenever we needed any kind of optical sensor or telescope.

The Ministry of Aviation, of course, was represented. There was the Ministry of Machine Building, which didn't even specify what kind of machines. I had heard said that their toys consisted of powder and munitions. I could never resist a somewhat sad observation. According to the number of appropriate ministries, Machine Building in general was almost officially considered the province of the military establishment. What gloomy symbolism, I thought. It sounded like the whole nation had only one overwhelming task: the building of a huge military machine!

For every topic on the agenda the first speaker, according to ritual, had to be from the enterprise or mailbox that represented the principal contractor for the project. Then the ball would be passed to the subcontractors. Within a few minutes, the ministers and the chairman of the commission had a chance to assess the state of every individual project. The main task usually was to identify who was responsible for being behind schedule. The poor subcontractors were usually immediately singled out—invited, as we would joke, onto the carpet to become the subject of arm twisting and other kinds of more sophisticated pressure. Apparently, that was what Georgy Petrov had complained about. He was the one most often called on the carpet.

The whole purpose of "the execution," as my friend Petrov called it, was to make sure the edifying lesson influenced "the boy to be beaten." "Press, press, press" was the motto and principal technique of the military-industrial establishment. It was presumed that an unlucky director or chief designer, after returning to his own office at the enterprise or mailbox, would transmit that procedure to the lower echelons. The sharp tongues immediately found the main algorithm, based on force and oppression. They paraphrased the famous dictum of dialectical materialism that forms the basis of Marxism, "the being precedes the consciousness," as "the beating precedes the consciousness."

When the meeting ended, my very first baptism in combat was over. I was much more fortunate than anyone could expect. However, later I realized how much I would have preferred to become a sacrificial lamb myself than to witness the collapse of the whole Mars project, a tragedy that came about due to the unexpected failure of a huge number of microchips and transistors that had been produced by the Ministry of Radio Industry.

In all planetary exploration, the supreme timetable from above dictates the rules even for such powerful bodies as the VPK in the Kremlin. No wonder our contractors within the Magnificent Nine were usually less than enthusiastic about taking part in flying to the planets. In their more or less routine work—producing tanks and rockets, often with enormous overkill—they knew they always had a last chance for maneuver. If they slipped off schedule, the sky would not fall on them. A few more arm twistings on the carpet—that's all they risked. But as tempered and experienced as they were, they knew they would survive the execution.

When the organization of a multiagency, multidisciplinary expensive project in the system had to overcome the lack of genuine stimulus or interest, only a firm hand—a heavy fist—could force the huge infrastructure (we called it the "cooperation") to accept these contracts. This was often necessary, given the dread with which the military-industrial complex regarded the space community's projects.

The ritual of acceptance, or the technique to make the contractors take responsibility, was very well established—a fact that I learned rather quickly. The idea for a new project, whether related to final procurement of new military technology or not, first had to be decided upon in a narrow circle of chief designers and their potential customers. This involved the representatives of various chief commanders among the military (like the air force, the air defense, or the navy). In the next round of lobbying, the authors had to win allies among wider circles of people in key agencies: the Defense Department of the Central Committee, VPK, and the ministries. Before very recent changes in the legislative structure of our society, I never heard of any need to lobby for the actual budget. The State Planning Committee (Gosplan) was automatically considered a kind of junior relative of the VPK family.

Only at the end, when the scale of operation and timetable were approved, was the final budget to support the cooperation drafted by a special department of Gosplan. We called this department the "Tenth Floor," the secret place behind the closed doors in the main building of Gosplan. But long before that final, more or less formal stage, the most difficult operation usually was to "bend" the restives—a huge number of industrial enterprises and mailboxes who were not impressed by a new bright idea, mainly because they were not ready to accept additional responsibilities and burdens above those already established by previous obligations.

It was then that the intimidating apparatus of the Central Committee, VPK, and regional and local party bosses was switched on to corner the dissenters and make them a final offer they could not refuse.

Besides the stick, there were also carrots. The system invented and introduced a multitude of bonuses and privileges. The least of them was the bestowal of medals of different grades of moral honor. For the hundreds and thousands of presumably heroic workers and engineers, the government established bonuses from the very beginning. In some ways it was reminiscent of the old Russian tradition of the titled lord expressing gratitude for his serfs or workers by rolling out the barrels of vodka. I had heard legends about Korolev, who in the beginning of the space age stimulated the team at the Baikonur launch site with the promise of barrels of spirit.

The last but not the least important dish on the carrot menu was the initially prescribed allocation of special resources to enlarge the size of a mailbox. This is how individual mailboxes gradually strengthened and expanded their estates.

This part of the deal apparently was called "the hay and the straw." Where such a definition of the bonus came from always piqued my curiosity, which was later satisfied by the old-timers of the military-industrial establishment: "This is the soft bed to be prepared for all of us in the event of a disastrous flop."

In general, the bonus system was a grand bargain that drove the military budget without any hint of a market economy and without caring about the tolerable limits of the national economy. The complex was like a huge black hole that sucked in most of the national resources. While the space program's part of the budget was only a tiny fraction, the share for science, which was represented by my institute and the Academy of Sciences, was indeed a drop in the sea. This provided the moral consolation that we used to justify our growing appetite for space research.

Over the years, government officials became more experienced and started to consider scarce resources before promising valuable assets as bonuses. To the great disappointment of contractors, we often found that the final decree issued by the Central Committee and Council of Ministers legislated substantially fewer trophies than were promised at the beginning of the project. By the eighties, Brezhnev had been mentally debilitated for years and the Politburo could no longer dismiss the notion of the forthcoming economic collapse of the country. By that time, Ustinov had changed into the military uniform of a marshal, appropriate to his new post as defense minister. He strengthened his position in the Politburo and became almost untouchable. It was then he uttered a sacramental phrase that never appeared in the open press but was delivered through the military-industrial grapevine: "Our people are very patient and tolerant. The only thing the country needs for its survival is bread and defense." While Ustinov did not explicitly include the circus in the formula of things required to make people happy, I am sure he counted on the continuation of his beloved shows in space.

A system that has no checks and balances—that does not even try to establish the limits—is doomed. But I believe the concentration of power in the hands of Ustinov, industry, and the military accelerated the final downfall significantly.

Ustinov was fond of astronomy. He visited from time to time the construction sites of the biggest telescopes. Gorbachev, before he came to Moscow in 1979 as a nonvoting member of the Politburo, was party secretary of the Stavropol region. Several times, Ustinov's sudden visits to astronomic sites were conducted in Gorbachev's company. Thus, Ustinov might have had a good chance of making a case for science before Gorbachev, had he known that the latter would one day be general secretary.

I never had much success with Ustinov, however. Perhaps my personal failure can be explained. Once and only once, at the very beginning of my space career, I had to ask Ustinov for a favor on behalf of my institute. There was a special telephone that was part of a privileged network called *kremlevka.* To have the kremlevka phone in your office was doubly important. Not only was it the chance to have immediate telephone access to the big bosses, but it was also a very important symbol of nobility.

My colleagues would ask me, "What is the phone number of your *kremlevka*? I couldn't find it in the directory." The network had a special secret directory.

"Sorry, I don't have a *kremlevka,*" I would reply.

Then my colleagues would exclaim, "This is impossible. It is so important for you to have it. Why didn't you ask for it?"

In fact I had asked. That was the problem. My predecessor Petrov had been given temporary status for such a phone for the period of the preparation and implementation of the *Apollo-Soyuz* project because of the role our institute played with the Americans. The bosses at the top needed direct access to information about this important project at any moment. After *Apollo-Soyuz* was blessed at the Nixon-Brezhnev summit of 1972, I inherited this telephone with the chair of director.

Frankly speaking, I never misused my access to it. I used it only for real business. After the project was over and the dust and the sprays of champagne had settled, I discovered to my disappointment that the *kremlevka* was to be disconnected.

After only two years in the office of director, I did not know what to do. My deputy, General Narimanov, who was considered a true courtier and knew about these things, advised me to make a direct phone call to Ustinov to ask for his intervention. Ustinov himself answered my call. After my explanation, he clearly was not enthusiastic. Thus, I was not particularly surprised when a few days later the *kremlevka* phone was taken away and the lines were cut. It was for "club members"—the upper echelon of the *nomenklatura*.[3] I was a scientist, and in their estimation a "white crow."[4]

My office never had a *kremlevka* again, at least not until I left the chair of director. What a consolation to learn that it was at last again installed! This good news came to me in the summer of 1990. The bad news was that I could not use the wonderful gift even theoretically—by that time I had remarried and my new wife and I were living in Bethesda, Maryland. The "honor" shone like the light of a dead star.

I had no better luck when I had to ask Smirnov for a small favor. It happened almost at the beginning of the *Apollo-Soyuz* project. Rumor was spreading that the government, in an effort to stimulate and support the participants in the project from the Soviet side, was distributing hard currency to purchase a few important pieces of hardware and some accessories. My assistants at the institute advised me to use the opportunity to ask for a small sum of this money to buy photocopying machines and a few decent printers. I thought that they were perfectly right since our institute had been given the role of providing the window dressing for the American visitors, who naturally would need to copy and print a lot of documents.

3. The party elite.
4. In Russian, this is the equivalent of a black sheep.

At one of those VPK gatherings at the Kremlin, I was asked to report on the state of readiness at the Space Research Institute before the critical stages of the *Apollo-Soyuz* project. Upon completion of my brief account, I added: "We urgently need a few small gadgets."

Smirnov obviously did not like my choice of words. He left his chair and slowly approached me. Standing next to me, he launched into a long sermon: "You are addressing the wrong man. Do you think I am sitting here as your commercial representative, as the head of the purchasing department for all of you?" The message was very clear. And though it was given to edify not only me but everyone, I was chosen as the freshman—the provincial—to be the whipping boy.

In Smirnov's defense, however, I must confess that he was not a rancorous character. During the many years of my space career I saw Smirnov a lot. Even if he was not particularly disposed to be friendly, he clearly had no antipathy toward me.

To be philosophical for a moment, the big bosses had a different formula for scientists. In one way, they considered us impudent beggars, like those described two centuries ago by Jonathan Swift in *Gulliver's Travels*. In Swift's story the hero was permitted to see the grand Academy of Lagado. The very first intellectual he met "was of a meager aspect, with sooty hands and face. I made him a small present, for my lord had furnished me with money on purpose, because he knew their practice of begging from all who went to see them."

For me, it was much more. I found it almost indescribably painful to see the indifference—even the denial of support—for interesting scientific space projects. It was difficult after such rebuttals to contain my anger and embarrassment. The old-timers of the military-industrial complex later taught me the formula that worked in our society. "Roald, take it easy. You have absolutely no idea why government is supporting your damned space science. Ustinov, Smirnov—they have absolutely no obligation at all. When they do it, it is out of sheer philanthropy. Be thankful for that. It is their beloved hobby, after they have fulfilled their regular, more important, duty of spending the taxpayers' money."

There was an element of truth in this interpretation. However, I learned from personal experience that these inveterate politicians did not simply support science as a hobby. Every small event in the implementation of a space science project, like a launch to the planets or a docking in space, was attended by every important person from VPK. Very often, Ustinov himself was present. I always wondered what was in it for them. To attend such an event required hours and hours of sitting in the control room, long after midnight, watching the display screens, or even, during the earlier years of the space era, flying to distant launch sites. Such a waste of time! And sometimes it was even risky.[5]

However, I discovered that their motivation was really quite simple. Late at night after a successful launching, landing, or docking, Ustinov or Smirnov would approach the *kremlevka* phone and dial: "Dear Leonid Ilyich, I am happy to report that the launch was successful."

That was the utmost happiness; that was the final reward. After a short conversation, the boss would have extended his overwhelming joy to the attendees: "Leonid Ilyich Brezhnev sends his heartiest congratulations to all of the participants."

That was the secret. These spectacular events were nothing more than an efficient instrument of politics for them, a way to be noticed and promoted. It was much easier to get a promotion in that way than by calling Brezhnev upon completing the opposite season of harvest: "Dear Leonid Ilyich, unfortunately this year's weather conditions were unfavorable."

After a launch, no matter how late it was, all the VIPs, as well as a few ordinary mortal engineers and scientists who had contributed to the success of the project, were invited for what was called a "tea party." I quickly discovered that at such gatherings, "platonic tea" was not the most important liquid. I usually found such drinking parties boring, so for a while I tried to leave as soon as possible. After several clearly unforgivable departures, my new friends from VPK warned me that my absences had been called into question—they could have said that "those who are not with us are against us." Such a notion was deeply ingrained in our psyche. It was, after all, one of the Stalinist axioms that had survived since my childhood.

5. In 1960, Marshal Mitrophan Nedelin fell victim to such an addiction. While the booster was in preparation for the launch, he ordered that his chair be placed in close proximity to the launchpad. The professional rocketeers and chief designers knew that it was dangerous and thus strictly forbidden. No one, however, wanted to countermand his orders. As fate would have it, the rocket blew up on the launchpad and with it the lives of more than eighty onlookers, including Nedelin.

But I had violated an even more important given. At such parties every toast that was offered had a special significance. My friends said that these toasts were how cooperation was decided and the future direction of the program was set. Life is the best teacher. Thus, I learned very quickly not only to attend such parties, but also to sponsor them. Nonetheless, for some of my fellow scientists and me, to attend these tea parties gave painful meaning to the notion that "science demands sacrifice."

Soon after the expansion of the international component at IKI, I got from time to time phone calls from our supervisors, telling me that they had an urgent need to meet with me at the institute to discuss important business at a tea party. I immediately understood that their interest was in my scotch. Because the institute's international contacts had enriched the space program with imported drinks, there was an immediate increase in the attractiveness of IKI and support for its projects.

Rather soon after the beginning of my career as director, I had to admit to myself, as painful as it was, that I had failed to follow the most important advice of Kapitsa and Budker. I had not achieved the status of "most favored director" for myself and "most favored institute" for IKI with the higher echelons of power. This was partly because I was always a poor letter writer and probably insufficiently eloquent verbally. As consolation, my only chance was to use Budker's recommended contingency plan: after giving up the siege of the big bosses, turn your attention to another target and establish friendly relations with the second echelon of power. Ingratiate yourself with the deputies, aides, and assistants—the whole group of modest but competent people who are rather invisible.

If the institute succeeded during my tenure, at least in a number of important scientific projects, it was because of the great contribution that was made by my "helpers" among the bureaucracy. They would tell me: "Don't overspend your time trying to impress the big bosses. Leave that job to us. Let's sit down and discuss every detail so we will be able to represent your interests. Then trust us—everything will take place according to our scenario." The good Lord knows, I trusted them.

Probably the most important minister whose attitude was critical to the institute's fortune in acquiring new spacecraft and launches was Sergei Afanasiev, the czar of the General Machine Building Industry. He was a huge man with large, sturdy hands. Like a "hammer striker," we said. When he chaired a meeting, the figure of this minister induced fear. His sentences for employees of the ministry, whether they were general designers or simple engineers, were brief and ruthless.

"You will get a serious black mark on your employees' record," he threatened.

I was told the man was vindictive and had a very long memory. Members of his team secretly referred to him as "the Big Hammer." Knowing this, I tried very hard to assure that our paths never crossed.

In 1975 there was a special meeting at Smirnov's office on the then-new field of remote sensing. Americans were getting a lot of publicity for their remote-sensing satellite *Landsat*, equipped with a multispectral scanner that for its time provided rather detailed images of Earth, including some over Soviet territory. The authorities were particularly concerned that the USSR was lagging behind in our own remote-sensing program. While analyzing the backwardness of our technology at the meeting, I said, "*Landsat* sends its data at the rate of almost twenty megabits per second. Unfortunately, information from our scanners is not digital at all. But even if it were converted, it would not be much more than one hundred kilobits per second."

I was unable to assess how much Smirnov, the head of VPK, was impressed by such comparisons. However, when we left the room through that famous door in the Kremlin, Afanasiev was clearly boiling.

"If ever again you bring up your damned megabytes, you will get into big trouble," he said to me, clenching his teeth.

Unfortunately, it was not a simple rhetorical threat. Rather soon, at the peak of the *Apollo-Soyuz* encounter—the most expensive handshake in the world—the Big Hammer and I both sat in the same part of the control room. Apparently, as I understood much later, at that very moment some of the IKI scientists were having a long phone conversation on an international line with their American colleagues. The discussion entailed trying to suggest a slight modification in the work of a very modest science instrument on board—the ultraviolet spectrometer. Someone probably told the minister that scientists had the courage to intervene in the flight program in process. With such an apparent challenge to his authority, the big man jumped up. Storming in my direction, Afanasiev prepared his fists as if he would use them like a big hammer. Then through his teeth he said, "Who gave you the right? I will squeeze your little balls."

I had no time to determine what was at issue before he was ready to simply knock me out. I spoke up immediately. "Don't you ever talk to me again with such a tone. I will not allow a minister or even a member of the Politburo to talk to me that way."

Everyone standing around us was genuinely stunned. Later on, Afanasiev's first deputy—an old-timer of the space program, Georgy Tulin—privately approached me with a word of consolation. "You are a brave boy. But you touched a very sensitive spot. My minister has daydreams of becoming a member of the Politburo."

Despite this obvious confrontation, time turned out to be the best medicine. The most remarkable outcome of this encounter was that eventually I established a rather tolerable working relationship with the Big Hammer. Without being afraid for what he had elegantly expressed as "my little balls," I invited Afanasiev to IKI. Over the succeeding years, he visited the institute several times and helped on some important projects. Although I don't think he changed his principal habits, in this particular case the Big Hammer was hammering for us, the scientists.

As I discovered much later, he apparently felt that he was an endangered species. While he was clearly the most successful and one of the most prominent Soviet ministers, he never received a further promotion. Not only did he fail to reach the level of the Politburo, he also never held the chair of deputy prime minister of the country.

The explanation was not found in the apparent rudeness and autocratic attitude he had toward his "subjects." After all, we were all products of the system, which was reflected in our instincts and conditioned by our reflexes. The explanation was probably hidden in facts that I learned only much later. His camp—or, better said, his mafia family—had for many years been in open conflict with the mafia led by Ustinov. Thus, Afanasiev was doomed. In 1983 he was relieved from his post in the space industry, and he had to leave the military-industrial complex forever.

For a while he was given the post of minister at the nonprivileged Ministry of Heavy Machine Building—a ministry of real machines, like excavators and cranes. However, nothing could help him in this new job. There were no bonuses or privileges as there were in the military industries. Furthermore, he had no expertise in these strange machines. His own hammer striking, unfortunately, didn't help him much either. Within a few years, he was forced to retire.

24 / The War of the Titans

➢

The ministers within the military-industrial establishment were the indisputable kings of their empire: they could appoint or fire almost anyone, including their deputies. Afanasiev had demonstrated that. Soon after the confrontation between Afanasiev and me at the *Apollo-Soyuz* docking, my generous counsellor-sympathizer Tulin was forced to resign. The minister had a very simple reason: Tulin belonged to a different family, a different clan, from that of rival Ustinov. It was unbearable for Afanasiev to have a clansman from another family in his fold. Afanasiev himself was a faithful ally of another powerful member of the Politburo, Marshal Andrei Grechko, the minister of defense.

Despite Afanasiev's power, however, there were a few people in the ministry that even the Big Hammer could not touch. A few leading designers of rockets and spacecraft, in the long and painful processes of strengthening their position, were often able to build their own feudal estates, declaring sovereignty even from their czar, the Big Hammer.

Korolev and his descendants were one of these estates. I can reconstruct the role and influence of Korolev only from stories and accounts I heard from my colleagues. He clearly was a mighty feudal lord, capable, it was said, of "opening the door of Khrushchev's office with his foot." If this were not the case, then there probably would not have been the first sputnik.

The only other person among the chief designers of the space age capable of competing with Korolev in influence, according to the legend, was academician Vladimir Chelomey. His name was much less known outside the military-industrial complex. Chelomey's successes were not in launching spectacular spacecraft so greatly needed by the government for propaganda, but in developing technology in classified areas, like the early generation of cruise missiles. Although these are low-flying objects, Chelomey's ambition always took him beyond the terrestrial atmosphere. Unfortunately, here he was perpetually behind Korolev, who enjoyed fame as the father of *Sputnik I* and the first manned flight in space.[1]

Chelomey was known only to a rather narrow circle of professionals. I met him in the early seventies, somewhat before my own career was launched into space orbit. Lev Artsimovich introduced me to him at Barvikha, a privileged Central Committee sanitorium, where I visited Artsimovich during his recuperation from a heart attack.

During one of my visits, he said to me, "You know I have a very unusual neighbor. I want you to meet him. He claims he is 'the most expensive man' in the Soviet Union. His name is Chelomey."

"What do you mean, 'the most expensive'?" I asked.

"That's in reference to the budget his mailbox has," Arsimovich said with humor. "In that sense he is, I am sure, a billionaire at the taxpayers' expense."

I did not see Chelomey for quite a while after that early encounter. But from the moment I joined the Space Research Institute, my new colleagues and subordinates indicated in their introductory course of anecdotal stories that Chelomey occupied an outstanding place within the Soviet space establishment. He had seemingly worked magic. During the Khrushchev era, he had had the wisdom and the luck to hire Sergei Khrushchev, the son of the country's premier.

My colleagues said that Chelomey was an absolute master at using their personal triangle for the advancement of his ambitions. It is not that Chelomey got hints or requests from Nikita Khrushchev to promote his son. The stories I heard gave the completely opposite scenario. It was Chelomey who had taken the initiative.

From time to time, smart Chelomey would make casual remarks, such as: "You know, Nikita Sergeyeivich, frankly speaking I was never expecting anything from Sergei Nikitovich that might have been outstanding—certainly not in such a short time. He is a smart fellow of course; we all know that. But I have found he is capable of real surprises. One day I asked him to give thought to one part of a project we were working on, which I myself considered almost unresolvable. As a matter of fact, I was quite skeptical about one man's capability of contributing

1. It was Korolev who persuaded the government to support a superlauncher for the Soviet counterpart to the *Apollo* missions to the moon.

significantly to this problem. I was wrong. I was really stunned at how quickly he has made a breakthrough."

What father would reject such a compliment focused on his offspring?

No one is capable of confirming how much truth there was in such stories. However, judging from the unusual collection of awards and decorations given to Sergei Khrushchev in a rather short period of his involvement in the Soviet rocket program, something must have been in those tales.

The employees at the mailbox joked that chief designer Chelomey enjoyed impressing the recipients of his messages with the final signatures: Chelomey, Khrushchev. However, according to the same stories, Sergei lost his spot in the privileged parking lot inside the perimeter of the installation the very day after his father was ousted from power.

With these stories in mind, I was psychologically well prepared when academician Chelomey invited me to his estate to talk about potential cooperation between IKI and his empire. In the autumn of 1973, still a very young director, I found myself making my way to the outskirts of Moscow, approaching a small town, Reutovo. The tightly guarded gates of Chelomey's installation were opened for my car, and I was directed to a parking place. There my black Volga joined the few other government cars of Chelomey's closest associates. In itself, it was a very important sign that the reception was going to be at the highest level.

I entered the huge room that served as his office. The room was in impeccable order and the souvenirs of his space successes lined the walls. Chelomey stood up and approached me in a very courteous way. "It is so remarkable that the legacy of Lev Andreevich is here with us in the space community," he said with a smile.

During the next few hours, I was exposed to the full force of Chelomey's eloquence and charm. I had been warned by my friends that I would play the role of audience in the theater of a single actor. They were right. He dominated the conversation, putting on a virtuoso performance.

Chelomey clearly wanted to impress me. I was introduced to a long series of mock-ups of his products, designed to fly everywhere in the atmosphere and above. From time to time, his assistants entered to hang more posters showing the color prints of objects currently in design or under discussion. He was rather disappointed, and at the same time intrigued, by the lack of admiration or enthusiasm coming from me. I skeptically viewed these unusual flights of fancy, materialized in this "most expensive" package for Soviet taxpayers.

Suddenly Chelomey sank into a chair, looking tired. "So, what do you want particularly from me? Do you need support?" he asked wearily.

I responded immediately, "Vladimir Nikolayevich, my institute dreams of being able, in the distant future, to launch a heavy and sophisticated astronomical telescope into orbit."

He jumped onto the subject immediately. "Why the distant future? I can already suggest to you a heavy launcher I have designed, tested, and introduced as a regular, almost commercially available booster." He was talking about the Proton launcher.[2]

"Moreover," Chelomey said, "I would soon be capable of providing a space home for your telescope in orbit."

"You mean your version of the orbital station *Almaz*?" I asked.

The literal translation of *Almaz* is "diamond," but this name was kept secret. Chelomey's orbital stations were launched under the name *Salut*.

"I don't think space astronomers in my institute would be excited about having their sophisticated and sensitive telescopes on a manned station," I ventured.

He laughed. "You are smart fellows. Don't worry. Currently I am converting this design into an unmanned heavy station based on robotics. Even my military clients are eager to have an unmanned version for their heavy instrumentation."

Now it was my turn to be intrigued and excited.

After that first visit, Chelomey and IKI were engaged in fruitful cooperation, designing an unmanned home for space astronomy. Chelomey as "the most expensive man in the Soviet Union" was also generous enough to financially support the early work on the scientific part of the project, specifically the astronomical hardware at IKI.

My association with Chelomey, however, was controversial, to say the least. Chelomey's name met with mixed reactions when I described my encounter with him to Keldysh a few days later. To my disappoint-

2. The Proton launcher was the largest Soviet rocket before *Energiya*, which is larger than the *Apollo* program's *Saturn 5*.

ment, Keldysh's reaction was cool. At most he promised to stay neutral in all my dealings with Chelomey. But I was disturbed by his reticence. Back at IKI, I shared my concerns with General Narimanov, my guru in issues of space diplomacy. His face immediately lightened up. In a conspiratorial voice he said, "Of course, I can explain everything to you. There has been a great war between the titans." This is a summary of his account.

In the 1960s at the peak of the early debate over the validity of antiballistic missile (ABM) defense, the strategic thinkers on both sides of the ocean suggested that intercontinental ballistic missiles (ICBMs), capable of delivering several warheads at once, could be much more difficult targets for an ABM to intercept. This perceived efficiency created a new style of arms race, this time for so-called MIRVed ICBMs.

The Soviet side had always relied on heavy boosters, so there were no insurmountable problems in designing multiple warhead systems. Sensing the rich prospects for awards, decorations, and bonuses, two parallel teams of chief designers suggested their own versions of the MIRV rockets. One of the two projects was promoted by the successors of the huge empire of Sergei Korolev and Mikhail Yangel.[3] The final result of their activity is famous, known by the name given by American experts, the SS-18.

At the same time, Chelomey led a team of competitors. Their version, too, finally materialized. It is now called the SS-19. How naive were those who thought that the arms race was a competition and an escalation of efforts between two or more adversaries on the international scene! The story of the SS-18 and SS-19 was a demonstration of an even more dramatic competition in the arms race, within one and the same ministry.[4]

The Big Hammer, Minister Afanasiev, was unable to make the final choice in favor of Chelomey—though he clearly was on his side. The reason was very simple: the SS-18, and the feudal lords promoting it,

3. Another early hero of Soviet rocketry, the founder of the Dnepropetrovsk enterprise.

4. Clearly Trofym Lysenko was wrong when he had tried to prove that there is nothing like an intraspecies struggle. We saw signs of it every day, in the most dramatic and expensive ways.

were supported by Ustinov, who was senior to Afanasiev. But Ustinov, in his turn, was also unable to make the final decision in favor of the SS-18. On the other side of the table, behind Chelomey and Afanasiev, there was a very powerful figure, the minister of defense, Marshal Grechko. The president of the academy and at the same time the chief theorist of cosmonautics, Keldysh, chose to support the SS-18 and Ustinov. Thus, the Politburo had to become the supreme judge.

Brezhnev, however was a softy when it came to dealing with military-industrial issues. Instead of making the critically important principal decision, he chose to compromise. Both projects were adopted in parallel.

There was no winner but rather a draw in that war of the titans. However, there was a loser: the Soviet people, who were forced to pay for both projects, without knowing anything about it. This type of resolution did not increase the security of the country, nor did it assure the better strategic stability of the nuclear configuration between the two superpowers. And it only created a puzzle for American strategists, who discovered the duplication from their reconnaissance data.

When I met Chelomey in my capacity as director of IKI, the echo of that dramatic war could still be heard. But both camps were capable of deterring each other in a seemingly stable strategic configuration—like that between two heavy cannons. In this case the cannons were Ustinov and Marshal Grechko.

The actual production and installment of the SS-19 was too dull a routine duty for the ingenious and creative Chelomey. He was already completely absorbed with another project, the story of which goes back to the time of Khrushchev, who had a fixation with U.S. aircraft carriers. Something, he thought, had to be done to restore the balance of forces on the seas. However, a straightforward symmetric response based simply on building Soviet counterparts to these dangerous vessels would have required an astronomical financial commitment and a great deal of time. These arguments, however, did not stop the Soviet military-industrial complex from investing in a tremendous naval buildup.

To address this, it was thought that something should be developed in space. At that time, even Americans, who perceived themselves to be lagging behind the Soviet Union in space technology, were discussing a project called the MOL (military orbital laboratory). Chelomey was excited by the concept. This is precisely what we needed, too, he thought. We should build a heavy spacecraft—a station—filled with sophisticated reconnaissance equipment, including radar that could easily locate aircraft carriers. These huge chunks of metal were ideal targets for radar.

The story I remember is based on Chelomey's own account. In choosing the size of his future orbital station for military clients, no sophisticated arguments could help him, he told me. To build, assemble, and launch the orbital station required only practical considerations. The body of the station, eventually to be built and assembled at his installation near Moscow, would have to be finally shipped to the Cosmodrome. "The maximum width permissible for railroad transportation is precisely four meters," Chelomey said to me with a wise smile. That was how Chelomey came to the maximum size supported by the practiced limits for railway transportation.

The MOL project on the American side never came to this stage of materialization. No matter—this particular idea triggered Chelomey's thinking and eventually led to the launching of the orbital station *Almaz*. Such are the dialectics of the arms race.

In the early seventies, Chelomey's installation was ready to launch the first military cosmonauts into orbit to work on board that very station *Almaz*. The authorities had to decide how to present such a purely military mission. The international public and the Soviet people already supported the principle that anything related to manned flight should represent a humanitarian mission. Should we now confess openly that the men we were sending into orbit were spying? No one would take such responsibility. So, *Almaz* was given the cover name *Salut 2*, after the orbital station that had already been launched a couple of years earlier by Chelomey's archrival—the Korolev Design Bureau, led at that time by Mishin.

This cover name was a real humiliation for Chelomey. He told me that even the very idea of the orbital station, with the sacramental diameter of four meters, had been stolen from him by Mishin and his team. Now to accept the name *Salut*, the trademark of his great rival, Chelomey regarded as the ultimate insult.

The launching of *Salut 2* (*Almaz 1*) met with bad luck. It was in flight for three weeks, but the crew was never sent because it was discovered that the station was losing air pressure. No one was able to explain whether it was due to a technical fault or an encounter with a meteorite. Chelomey, of course, preferred the second explanation. There were two subsequent launches: *Salut 3* in 1974 and *Salut 5* in 1976, which had better luck. They operated in orbit for up to one year and were able to accommodate three duets of military cosmonauts.[5]

However, the legendary war of the titans found a new dimension in space, this time in launching orbital stations. During a period of slightly more than ten years, seven Soviet space stations under the name *Salut* were launched into orbit. Three of them were of Chelomey's design, and the remaining four were launched by the successors to Korolev. The latter, as a matter of fact, had much better luck. Their instrumentation was much more diverse, and they did not try to keep everything under the control of the military.

Valentin Glushko, who assumed the leadership of the former Korolev Design Bureau, at once had the strength to compete with Chelomey. Both feudal lords suggested a wide spectrum of new projects, ranging from orbital stations to new launchers, as well as their own versions of the future Soviet counterpart to the shuttle spacecraft. But the outcome of that great war was decided not by the talents of the engineers and designers working on their blueprints, but within the political corridors of power.

Glushko staked everything on his relationship with Ustinov. In his literary activity—writing articles and books mainly glorifying his own contribution to the Soviet space program—he found room to paint Ustinov as a generous and wise supervisor. The second invaluable asset that Glushko had was the support of another member of the Politburo, at that time even more important than Ustinov: the secretary of the Central Committee, Andrei Kirilenko. Glushko had inherited Kirilenko's son-in-law Yuri Semenov, a young engineer at the Energiya Design Bureau.[6] Glushko quickly started to promote him. He sensed from the beginning that manned flights were going to bring much more important dividends than any other complicated unmanned space undertaking.

5. Despite all precautions, it was obvious even for a layperson that these stations had military functions. Otherwise, no one could explain the timetable designed for the crew: to stay awake over America and sleep over Russia.

6. Sometimes interchangeably known as the Korolev Design Bureau.

Chelomey was Glushko's only competitor in the new space war, with his own version of the orbital station, *Almaz* and the Proton launcher. Again, the strategic balance was supported at the very top level of the Politburo by Marshal Grechko. As before, the Soviet taxpayers—as they had with the SS-18s and the SS-19s—had to support two parallel competing versions of the space station. They paid for these enormously expensive undertakings, again without knowing it, because both genetic lines of this station had the same official cover name, *Salut*. However, the outcome of this war of the titans came unexpectedly soon.

In early 1976, Marshal Grechko suddenly collapsed from a heart attack, leaving Chelomey unprotected. The lucky star of the once "most expensive man in the Soviet Union" went down. Kirilenko and Ustinov reacted immediately. A few weeks after Grechko's death, at the Twenty-fifth Party Congress of the CPSU, they seized the moment and promoted Glushko to the ruling body of the party—its Central Committee. From this moment onward, Glushko concentrated in his hands not only the power of an enormous space empire, but also the political power of a commissar, capable of overwhelming anyone in the space establishment.

The rest was predetermined. As chess players would comment on the undisputed advantage of one of the players, the remaining issue was defined by routine techniques. The very first move that Glushko made in that position was to deny Chelomey contract for his orbital station. The orbital station *Almaz*, whether it was almost completely military or partially civilian, was terminated as an essential part of the Soviet space program.

I remember visiting Chelomey at this very moment to talk about current issues of cooperation between IKI and his huge *Almaz* company. As inexperienced as I was in the issues of high military-industrial politics, I understood immediately when Chelomey softly declined to continue the conversation on the astronomical interests of IKI. Suddenly he began an interesting psychological exercise.

He said, "You think you can impress anyone in the government by talking about sending space telescopes into orbit? I don't think so. If you were to tell me you have an idea to fly in space an instrument that could sense the submarines submerged in the ocean, then I could immediately invite you to visit the prime minister. That would be a great hit."

I did not know at that time that Chelomey was feverishly looking for a miraculous salvation. Ustinov had just become the minister of defense. Alas, there were no more cards in Chelomey's hands. The new defense minister, soon to become a marshal, started to methodically strangle Chelomey. He annulled all the military contracts given to Chelomey's enterprise for space flights; he canceled even those that were scheduled in unmanned mode and originally requested by the military.[7]

The thirst to humiliate and punish Chelomey, his former adversary in rocket and space wars, was so overwhelming for Ustinov that he even signed an order to demolish and discard extremely expensive hardware that had been accumulated for final integration in this ill-fated orbital radar station. Chelomey and his company were put by Ustinov on virtually a starvation ration. The only supporter of Chelomey still in power was Big Hammer, Afanasiev. But he, too, was also suffering from the absence of protection in the Politburo.

After Chelomey had been stripped of his budget, he spent many hours in a small corner of his installation with a couple of assistants, trying to experiment with surface waves at the interface between two fluids.

"What do you think of my experiment?" He demanded that I answer. As much as I was touched by Chelomey's clear and genuine desire to recover his intellectual freedom by finding a small asylum within his huge multimillion-ruble military-industrial corporation, I couldn't help to create an illusion for him. I directly answered, "Vladimir Nikolayevich, Peter Leonidovich played with this type of toy in the late forties when he was under house arrest."

Chelomey felt disappointed. "Do you think my results don't represent anything interesting for nonlinear mechanics?" I couldn't even try to explain to Chelomey how far the field of nonlinear science had moved since the early years of his brief passion for pure intellectual pursuit, before he became overwhelmed with the desire to become a principal actor in military technology.

A few months later I got a phone call from Chelomey. He told me that he had completed his experiments and had sent the paper describing his experiments to the presidium of the Academy of Sciences, where they were going to be presented as a report. He wanted very much for me to come and participate in the discussion of his paper. I knew how impor-

7. These were, like the heavy unmanned satellite, essentially a robotic modification of the space station *Almaz*, equipped with sophisticated and sensitive radar for military reconnaissance from orbit.

tant it was for him to have a chance for such self-affirmation. After all, he had to show the academicians and some of the leaders of different military-industrial enterprises that he was not on his knees. This demonstration had to prove that his spirit was free and capable of keeping him intellectually alive. Unfortunately, I missed his talk because of an important engagement on the same day. From that time onward, he gradually faded from my horizon. I cannot now recall having any lengthy conversations with him after that.

My friend Yevgeny Velikhov told me a dramatic story, in which he was involved in another of Chelomey's attempts to make a comeback. In a manner reminiscent of Kapitsa's effort to attract Stalin's attention by inventing the technology for strategic defense against aircraft and missiles, Chelomey was driven by the same logic. He wrote a letter to Brezhnev—bypassing Ustinov, his enemy—in which he suggested a spectacular scenario capable, he thought, of saving the country from massive thermonuclear attack. His invention was based on placing hundreds of tiny interceptor rockets on space platforms, somewhat similar to the one described a few years later—in the early eighties—by General Daniel Graham in his book *High Frontiers*.

Chelomey's proposal was rejected. I don't think Ustinov had needed to bring his old animosities forward to kill this particular project. The referees and participants of peer review were smart people—among them, Velikhov and Khariton. The proposal was Chelomey's last desperate attempt to regain his power and role in the Soviet military-industrial complex.

25 / War Games in the Academy

➢

The interesting outcome of Chelomey's failed proposal was that he never asked me to participate in assessing and promoting his last invention. I think he knew from the beginning that I was a "white crow" in Soviet military-industrial circles. In that sense, he was completely right. IKI had had to fight, from the very beginning of my directorship, long and hard against continuous pressure from the military to accept contracts. In this fight, I also felt like a lone warrior. The presidents of the Academy of Sciences did not help me defend the Space Research Institute as the only place in the country dedicated to nonclassified open space science. They saw no importance in maintaining a strictly civilian program. But at least the ailing Keldysh, in his last year as the academy's president, did not pressure IKI to participate in the military space program.

The last serious joint battle I fought with Keldysh as my ally was the battle around the Soviet response to the American space shuttle. What should the Soviet position be to such a new space concept of reusable transportation? Many different scenarios had been suggested by a group of chief designers in the Ministry of General Machine Building and the Ministry of Aviation Industry.

One very popular point of view was that the Soviet Union would not need such reusable transportation, capable of launching a crew in space and, at the same time, carrying heavy payloads. This view was supported particularly by Chelomey and the air force. The scenario they suggested was to combine the reusable capability, as well as normal launches, with expendable single-use rockets. The crew could go up and down in a tiny version of the shuttle, which would be considered a jet fighter rather than a bomber, relative to the size of the original American shuttle.

However, Keldysh and I were supporters of a completely different view. From the very beginning of the Soviet debates on shuttle-type technology, there was a general understanding that economically reusable transport would be unable to compete with normal rockets. Even the overly optimistic forecasts suggested by early NASA reports on the shuttle project had been unable to outweigh the extremely low cost of expendable boosters.

Keldysh and I put together a workshop focusing on the opposition's counterarguments. Participants considered many aspects of the shuttle, including its possible use for complicated maneuvers in orbit. Even here, we decided the shuttle could provide nothing new: with a substantial amount of fuel and engine power, any spacecraft could perform such maneuvers. Then someone suggested a scenario in which the shuttle could make deep dives in the atmosphere up to an altitude of 150 kilometers. But what for? What would be the purpose of such diving maneuvers? To come closer to a potential target? To take a photograph or a radar image? Or to be ready for a final doomsday suicidal attack?

Though we tried very hard, the workshop was unable to find even one single scenario in which the shuttle could provide a comparative advantage. Finally, I drafted a negative response to the government's request for the Academy of Sciences' opinion, in which I stated that the academy did not see any sensible way to use this Russian version of the shuttle. Cautious Keldysh, however, did not want to get into conflict with the military, so he modified my wording, saying, "We do not see any sensible scenario that would support the shuttle for scientific uses."

More or less pessimistic assessments came from almost every corner, except maybe from one group of chief designers in the aviation industry who were interested in getting the huge and exciting contract. Despite our assessment, however, the final decision was to implement the program. I heard that it was adopted mainly due to insistence from Ustinov, who had made the following argument: if our scientists and engineers do not see any specific use of this technology now, we should not forget that the Americans are very pragmatic and very smart. Since they have invested a tremendous amount of money in such a project, they can obviously see some useful scenarios that are still unseen from Soviet eyes. The Soviet Union should develop such a technology, so that it won't be taken by surprise in the future.

The compromise version promoted by Chelomey and the air force was also rejected. Furthermore, the government finally decided that the Soviet shuttle, *Buran*, had to be heavier than the original American version. In the event of any unpredictable development, we would always have the edge over the Americans.

This debate was the last serious battle—on a governmental level—in which Keldysh participated in his capacity as president of the Academy of Sciences. The next president, academician Anatoly Alexandrov, came from a different corner of the military-industrial complex. He brought with him a corresponding approach to such matters.

Keldysh retired from his post as president of the academy and Alexandrov assumed that position in late 1975. I knew Alexandrov from my years at the Kurchatov Institute, although I never understood what his particular duties and responsibilities had been at the institute. Apparently, he was Kurchatov's confidant. Coming from the same school of Leningrad physicists under academician Abram Ioffe, Alexandrov, like Kurchatov, changed his area of interest within physics a few times before finally coming to the nuclear program.

At the beginning of World War II, Alexandrov was the first to realize that physicists could contribute to strengthening the national defense. He went to the navy and launched a program that demagnetized the steel-made bodies of warships to make them invulnerable against ocean mines navigated by magnetic field–sensitive sensors. As a matter of fact, for a while young Kurchatov was Alexandrov's assistant at this job.

Probably because of these years and this experience, Alexandrov fell in love with the navy, and with the sea. Whatever he did in the future, he always displayed interest in the military—and specifically in naval programs. He was always surrounded by admirals and sometimes generals. Clearly they were the most respected species in his life. In that passion, he paralleled maybe only Leonid Brezhnev.[1]

After the war, Alexandrov was Kurchatov's companion at his nuclear installation. During the most difficult years with Beria, it was Alex-

1. During the years of World War II, Brezhnev himself was a politcommissar at different places on the front, advising generals on how to keep the morale of the army as high as possible. Since that time, Brezhnev had developed a deep respect for the generals. Years later this attitude could be felt in the Kremlin by the very way he ran the defense program and supported the military budget, which gradually dragged the nation to complete economic bankruptcy. In retrospect, it was unfortunate for the Soviet taxpayer that both of these men held the highest ranking positions in their fields. The influence of the military—and the navy—would now be felt all the more strongly.

androv who was assigned to take over Kapitsa's institute while he was under house arrest. This assignment has always been considered the trickiest and most controversial part of Alexandrov's career. Despite this, however, I never sensed a deep animosity in Kapitsa's attitude toward Alexandrov after Kapitsa was brought back to active life. My instinct tells me that Alexandrov tried his best not to alienate the disfavored academician.

Jointly with Kurchatov, Alexandrov presumably had the courage to keep the legacy of Kapitsa in the institute alive. He did this even at the risk of being persecuted by Beria himself. After Kapitsa was brought back to the institute, Alexandrov returned to Kurchatov's comradely embrace and took the position as his deputy at the institute. That was when I first met him.

When Kurchatov died, no one even asked who would be his successor. Everyone took it as natural that it would be Alexandrov. Apparently, Alexandrov was quite a generous man, capable of building human relationships. And as one of Ioffe's pupils he had a sufficient background in general physics to lead discussions on different scientific and engineering subjects—even though he himself was not an outstanding expert in a particular field.

Since Alexandrov was never considered especially bright, his nomination as president of the Academy of Sciences came as a surprise to many of us. Rumor told us that such a promotion came as a result of intense lobbying behind closed doors. It was said that the most active part was played by Yuli Khariton, who cared most for the future of the nuclear establishment to which Alexandrov belonged.

Apparently Keldysh eventually supported Alexandrov's candidacy. I believe that Keldysh's sentiments in favor of Alexandrov were motivated by his own deep disappointment in the rise of bureaucracy in the Academy—the very bureaucracy he himself had so much strengthened. Alexandrov—with his informal manner of rejecting any kind of written draft, even for the most important speeches at party congresses—was the personification of such an antibureaucratic stand.

When Alexandrov was finally appointed president, Keldysh invited me to pay a joint visit. We had a rather long and friendly conversation. Alexandrov frankly confessed from the very beginning that he knew very little about space. If he had ever supported the Soviet space program, he said, it had been in the form of toasts raised for its success. (Alexandrov, in private circumstances, was quite fond of such partying and toasts.)

Later it became clear that Alexandrov was not going to change his positive, if passive, attitude toward the space program. Moreover, he was not going to pay more attention to its management in the Academy of Sciences than he had before coming to its president.

As friendly and unbureaucratic as the new president was toward the scientists in the academy, he was even friendlier to admirals and generals. Crowds of them soon overtook the headquarters of the academy's presidium. The admirals and the academicians quickly learned that the old love and passion of Alexandrov—"naval games"—was going to be brought to his new headquarters. Here this passion would be joined with "academicians' games" to create a new genre of applied science, sponsored in the interests of the military-industrial complex. This became clear to us very soon.

If, during Keldysh's time, the Interagency Council on Scientific and Technical Aspects of Space Exploration had been the favorite advisory body for the president of the academy, Alexandrov established his own favorite child: the Council on Hydrophysical Problems. This program was oriented to promote the interests of the navy. The aura around space science was brushed away by the smart ones and the sycophants who rushed to join the new scientific council and the program.

Even those who were originally considered space scientists and engineers could not resist being dragged away by the gigantic planet-size tidal wave sweeping them toward naval research. The most beloved topic of the military became the search for the American fleet of submarines. My colleagues in space exploration hurried to suggest the importance of placing special magic sensors on spacecraft that would detect submarines from orbit, using microwave technology and its ultimate materialization: synthetic aperture radar.

The temptation to join Alexandrov's new program was irresistible even for some of my own collaborators at IKI. That's why I had to fight for several years the drive to convert the institute into a kind of supplement to the military-industrial complex. I was engaged in hot and bitter debates at the scientific council of both the institute and the academy. If IKI were to begin to take military contracts from the complex, I argued, who would be the winner? The military-industrial complex in the country would get one more, but very tiny, secret mailbox. But fundamental space science, with its prospects of international cooperation, would lose its only dedicated asset.

My friends among the clerks at the military-industrial commission told me once what was said in their circles about the director of IKI: "He is exterminating even the remnants of military-oriented research with a hot iron sword." I felt proud of such an assessment; I took it as appreciation of my efforts. But, frankly speaking, I was far from reaching a complete and undisputed victory. My subordinates, enthusiastic for contracts with the Ministry of Defense, from time to time still pushed and lobbied, especially in areas close to the interests of President Alexandrov.

This was the simple and straightforward thinking of my colleagues: every submarine, independent of how deeply it is submerged, is moving in the ocean. As it travels, it adds to the waves already existing in the natural environment. After several cascades of conversion, some of these waves could, in principle, directly or indirectly reach the ocean surface and leave an imprint on the overall wave pattern, which then could be detected by spacecraft. If equipped with appropriate sensors, such space observation platforms should be able to identify the presence of submerged submarines and even indicate their precise location. If one day the actual disposition of every nuclear-powered submarine in the ocean were known to military planners, the naval submarine component of the strategic nuclear forces of nuclear deterrence would lose its uniqueness—its perceived invulnerability. That was the daydream of admirals and academicians.

For a number of years after Alexandrov became president of the academy in 1975, the most desired and respected guest in the academy was Admiral Sergei Gorshkov, the commander in chief of the Soviet navy—famous for his success in the enormous military buildup on the seas. Alexandrov himself was particularly fond of Admiral Gorshkov, and I am sure the admiral received a record number of the president's toasts at different "tea parties." The sentiments Alexandrov expressed in favor of his naval friend exceeded all other toasts, particularly those in support of the space program, which undoubtedly trailed far behind.

We members of the academy were so accustomed to hearing the name of Gorshkov that one day, when we heard the rumor that there would be a special chair at the academy elections for Admiral Gorshkov in recognition of his "contribution" to oceanography, no one was surprised. Instead, most of us were rather curious about how the president of the academy was going to be able to persuade his sometimes disobedient academicians to elect the commander of the navy as their comrade-scientist.

This was a special period at the end of the seventies. Big bosses in different offices—in the Council of Ministers, the Central Committee, and Gosplan—sensed that, with Alexandrov and the changing mentality of the academy, the bosses had a great chance of getting a certificate of nobility—of getting elected as a member of the academy.

The Academy of Sciences prepared for the next electoral campaign. Several VIPs, including ministers, were to be nominated as candidates in the elections. Admiral Gorshkov was the most famous name among them. The normal academicians were told not to worry; there would be no competition for this kind of chair. The government would open special vacancies for its representatives in the academy.

In many ways, the drive among very important persons in the government to become a member of the academy, reflected the general corruption of the late Brezhnev regime. It was as if everyone were trying to tear off a piece of pie, as if the days of the regime were numbered—like the last days of the Roman Empire. My friend, an academic old-timer, commented on this issue: "When an academician is old and inactive, he is still an academician. If you take an important governmental figure who is out of a job, what would he be called? A political dead body! This explains why all of these people are trying to get the regalia and privileges of the academicians."

At the very peak of expectation over the forthcoming academy elections, something unpredictable and unexplainable happened at the very top—at the level of the Politburo. One day we were told that a special order had been issued by this ruling body, decreeing that members of the *nomenklatura* could not run in academy elections without the special approval of the Politburo. We could only guess at the real explanation for this new rule. We tried to figure out which of the two plausible versions was most decisive in the adoption of such a decree.

One scenario was that, facing such failures as the one of Sergei Trapeznikov at the last election, the government did not want to be humiliated by having more important members rejected by disobedient and irresponsible academicians. The second version, which I personally liked more, hinted that the comrades-in-arms who sat in the Politburo and in the Council of Ministers felt genuine jealousy toward the luckiest of them. Imagine how they would feel if one of their rivals, on top of other political decorations and attributes of power, were to get academic honors, too!

Thus, Admiral Gorshkov, the czar of the oceans, lost his chance to add to his naval crown the hat of an academician.

The general drive for military applications during Alexandrov's reign made it almost unbearable for me to stay on as director of the Space Research Institute. I couldn't find support in the ruling body of the academy, or even a minimal expression of interest for what IKI was doing in space. Several times I tried to resign from my post. Once, I almost came close to succeeding. Aleksandr Prokhorov, my direct boss in the Department of General Physics, agreed to support my resignation. With his preliminary approval, I delivered the letter of resignation to President Alexandrov. I was hoping for final liberation.

In the academy's corridors the know-it-all types were talking about a couple of important contenders for the soon-to-be-vacant post of director at IKI. I don't know what kind of arguments were finally decisive, but both candidates did not get enough support, and I was put under very strong pressure to stay in my position, even if temporarily. Alexandrov promised me that I would be liberated from the day-to-day boring administrative routine, and I would be regarded primarily as the science director.

There was another factor, probably no less important in keeping me in that job. The pressure inside the institute was great. My colleagues and fellow scientists did not want me to resign. I pacified myself with the thought that I would remain only as a temporary assignment. But I must have been tranquillizing myself to think that that was possible. What was to be temporary continued for almost another nine years.

It is a sad irony, and the last posthumous salute to Chelomey, that the radar images obtained from his reanimated unmanned station *Almaz* ("Diamond"), are also no longer secret. Taken out of mothballs a few years after Ustinov and Chelomey died, it was launched into orbit in 1985. The miraculous comeback of *Almaz*, ten years after Marshal Ustinov ordered its demolition, is in itself worthy of being a Russian fairy tale.

Apparently Chelomey, with the silent support of the Big Hammer, did not implement Ustinov's order to destroy the *Almaz* hardware. Chelomey and his closest associates decided to keep the disassembled components of the valuable station carefully protected in a secret place, in one of the corners of their estate—perhaps waiting for better times to come. It was a bit of a paradox that at the very heart of the military-industrial complex, a huge stock of expensive hardware was kept like a form of military space samizdat. Chelomey's ingenious assistants obscured the hardware from the search of unwelcomed inspectors by surrounding the stored components with signs that read: DANGER—RADIATION.

Chelomey did not live to see the triumph of his radar on the *Almaz* station, and he was never able to overcome the disintegration of his once-biggest empire in the Soviet military-industrial complex. Most of his subcontractors—his vassals of a kind—were given independence and the right to build their own feudal estates. One of them, originated by Chelomey many years before as a response to Khrushchev's desire to keep an eye on American aircraft carriers, was eventually established as an independent mailbox, Red Star. It supplied space radar that could detect aircraft carriers with nuclear reactors as energy sources.

American strategic analysts named the spacecraft carrying such radar and nuclear reactors Rhorsats. I bet no outsider could explain what had been the purpose of launching heavy nuclear reactors in space. After all, each of them had barely three kilowatts of electric power.[2]

To justify assembling a reactor in space, even when it was not necessary, one had to have very strong political support, and the chief designer of Red Star, Vladimir Serbin, clearly enjoyed it. His father, Ivan, complemented the necessary critical mass with his heavyweight post as polit-commissar of the military-industrial complex in the Central Committee of the CPSU.

Until they were discontinued quite recently, altogether thirty-two Rhorsats were launched. A few times they kept alarmed terrestrials at risk. In 1980 a Rhorsat under the code name *Cosmos 845*, lost control and fell on northern Canada. It was an unintended and unexpected form of

2. Even less sophisticated contemporary non-nuclear technology is capable of supplying bigger amounts of energy from routine solar panels. But for nuclear reactors with even such modest electric power, it would require keeping in orbit a critical mass of uranium 235. Such an unjustifiable waste of precious nuclear fuel, which enhances the risk of disaster, especially at launch, is bewildering to environmentalists.

export of Soviet nuclear technology—for which Canada, as a penalty, fined the Soviet Union $2 million.

Most of the family of Rhorsats were eventually relaunched to higher orbits, where they are capable of staying practically forever as a circling cemetery of nuclear reactors, 600 kilometers above the earth's surface. They remain an eternal monument to the inner logic that once drove the Soviet military-industrial complex.

26 / From Here to Eternity

With Chelomey stripped of his power and enormous budget, Valentin Glushko was left without a competitor. Now *he* was free to think about eternity. He had already been decorated with every imaginable award. His "trophies" included five Orders of Lenin, a Lenin Prize, several State Prizes, and an innumerable quantity of the medals of slightly lower value. However, the most important award—the one that could ensure its recipient a place in perpetuity—was a double Hero of Socialist Labor. By charter, the lucky awardee was entitled to be glorified, at government expense, with a bust built at his or her birthplace.

Glushko was born to make the city of Odessa happy. How lucky this most famous port on the Black Sea was, founded at the end of eighteenth century by Catherine the Great. The discovery of the site on which the city was built was an object of special pride for Count Potemkin-Tavrichesky.[1] However, to persuade the empress to come and give her blessing was no easier than defeating the Turks at this strategically important coastal site. To attract Catherine the Great, the victorious generalissimo had to display the talents of diplomat, courtier, and even architect. He erected those famous "Potemkin villages" along the road the empress took to the Black Sea.

Two centuries later, Glushko took his mission to Odessa no less seriously. He wanted to find the most appropriate place in Odessa to put his bust. The part of Odessa he finally chose is world famous. The mutiny on board the battleship *Potemkin*, the pride of the Russian navy on the Black Sea, had taken place when it was anchored in the harbor of Odessa. However, Odessa's real recognition came after a great director of early Soviet cinema, Sergei Eisenstein, filmed his masterpiece, *The Battleship Potemkin.*

1. The addition of the name Tavrichesky commemorated the conquest of the Crimean peninsula on behalf of the Russian crown.

Much earlier Odessa had known the young Aleksandr Pushkin, exiled by Czar Nicholas I to spend a couple of years here. The monument to Pushkin was built on the most prestigious and picturesque boulevard Prymorsky ("Seaside"), which wended along the high coast. The memorable long stone staircase in *The Battleship Potemkin* brought visitors from the port and beach directly to that famous street. This was the most beloved place for all Odessans to come and pay tribute, not only to the great poet, but also to the great architect and builder of Odessa—a foreigner, Duke Richelieu, a descendant of the noble French family famous for Cardinal Richelieu. The monument to the duke is just one block from Pushkin's.

Thus, my space colleague Glushko decided to join two other great heroes of Odessa and chose this very street for his own monument. Not all of his fellow countrymen, however, were happy with this decision. There were angry letters to local authorities and to the central government protesting the choice of location for his monument. However, Glushko was already a member of the Central Committee of the CPSU, so his influence in the country was much stronger than that of any of the leaders of the local government around this street in Odessa.

With that difficulty removed, Glushko discovered another problem: how to get a decent monument on the standard five thousand–ruble budget allocated by the government. At that time in the mid-seventies, that modest amount of money would probably have been enough to build a rather straightforward, unpretentious bust from simple cheap stone, appropriate for installation in a small or modest-size town. But it certainly was not enough for the man who considered himself equal or even superior to Korolev in the Soviet space program. On this lovely street of Odessa, his monument clearly had to be equal in quality and grandeur to the monuments to Pushkin and the duke. It was a matter of noblesse oblige, and as rumor had it at that time, academician Glushko decided to contribute his own money. As it turned out, even curious Odessans were unable to find out how many rubles were spent from the pocket of the general designer.

The monument itself was made from titanium, a rather expensive and very special metal used widely in space. Such a delivery clearly came courtesy of Glushko's own design bureau. The construction of the bust was finished on deadline, according to the discipline well accepted in the space industry.

After arduous political machinations by Glushko, everything was ready for the launch of an important and solemn opening ceremony. The hero, and at the same time the purchaser of his own bust, waited for an invitation from the local officials to attend the ceremony at the monument at which speeches glorifying him would be followed by loud applause. After all, there are probably very few people who are given the opportunity to stand next to their own monuments—to accept such a lifetime honor. Maybe only kings. But then, Glushko was a real space king, now unchallenged in his own country—"the homeland of the space age."

The monument itself was protected by a special cover in anticipation of the forthcoming celebration. In perfect Odessan humor, however, that was not to be the end of the story. The moment everything was ready, the Soviet government changed the very charter dealing with building busts for double (or more) heroes. The government's untimely decision was motivated by another famous hero, the one who had been decorated with the title Hero of Socialist Labor not twice but *three* times. It was Andrei Sakharov, who, according to the same charter, had for many years deserved to have his bust put on one of the streets of Moscow. Given Sakharov's political activities, however, it was clearly unacceptable for the government to promote the famous dissident, the fighter for human rights, in the capital of the country—close to the headquarters of the Communist party and the Kremlin's rulers.

To avoid this embarrassment, sly geniuses in Moscow changed the rules on busts. Under the new charter, only those double or triple heroes whose awards were presented for nonclassified achievements would deserve to have the bust erected in their lifetimes. Those who were decorated for their contributions to national security and for classified work now had to wait for the bust to be installed only posthumously.

Poor Glushko was given another headache and problem to overcome by courtesy of his fellow academician! He tried hard to get a special exclusion from the rule, in consideration of his substantial material investment in building his monument. Who knows what might have happened to his efforts had Sakharov not continued his ever-escalating war against the authorities. Finally, the government decided by special decree, to strip Sakharov of all of his decorations and prizes "for actions damaging the Soviet state."

After such a decree, there was no more need for the artificial charter depriving masses of simple and faithful Soviet double Heroes of Socialist Labor the chance to be appreciated and acknowledged by fellow countrymen in their lifetimes. Glushko was finally memorialized in the company of Pushkin and the duke.

Finally counted in that number with the other men of Soviet cosmonautics, Glushko started to take much greater interest in running the actual spaceflights, on behalf of his Energiya Design Bureau. He quickly sensed how strong a political instrument it is to send men into space, and he was intently looking for new initiatives to surpass previous achievements. His attention went first to the duration of spaceflights. He discovered that continuing to increase the record of the crew's time in orbit would be an inexhaustible resource for bringing the attention of important government officials to the space program and building a political power base for himself. With only a slight modification of the *Soyuz* space station, inherited from Korolev and Mishin, cosmonauts were sent into orbit under the sponsorship of Glushko. They increased the record of the crew's time in orbit up to almost one year.

IKI was not particularly fond of the manned flight program. However, in some ways we had to cooperate with it since it would have been political suicide for us to completely reject the chance to participate in the program. Thus, from time to time the Space Research Institute and the Academy of Sciences discussed what other kinds of scientific instruments or payloads could be designed and delivered for the cosmonauts to play with in space.

Frankly speaking, we tried to do our best, and it was not our fault that essentially nothing terribly exciting came out of such spaceflights. The explanation was rather simple: manned flight was accompanied by a long list of restrictions. One of the principal obstacles associated with sending a crew into space, at that time and still now, is caused by the necessity to maintain certain prescribed strict regulations. For every given kilo of weight brought into orbit for scientific experiments, many more kilos were spent for life support systems for cosmonauts. The very presence of a human being on board created severe restrictions for any payloads requiring high voltage. In fact, these were very often needed for important scientific experiments. The same was the case with chemical incompatibility with the local atmospheric environment inside the space station's modules. Nothing could be in violation of the safety regulations for cosmonauts. And it was quite understandable.

In addition, there was a problem with the inflexibility of the schedule. The itinerary of the crew in space is completely controlled by the need to survive. For instance, no scientific instrument could be left switched on when the crew was sleeping. Once, in IKI, we made statistics concerning the fraction of time in orbit when an average scientific instrument or payload could be operating. The big bosses of the military-industrial complex and the Ministry of General Machine Building were unpleasantly surprised when we gave them the final figure: about 2 percent of the flight time.

To keep the "right stuff" in orbit was an extremely inefficient way to do science. So, in our efforts to support the manned flight program, we provided rather simple and routine experiments that would otherwise be regarded as boring to most scientists. That is why there was no love affair between space science and the manned flight program.[2]

While the straightforward continuation of manned flights in space clearly produced political dividends for Glushko, he was desperately looking for new, politically attractive initiatives. One of the earliest among them, I believe, was his idea to revive women's flights in space. Almost fifteen years had passed since the only female cosmonaut, Valentina Tereshkova, had flown on one of the early, simplistic versions of manned spacecraft. Upon return to the earth, she was immediately promoted to chairperson of different important political committees. Despite this, her flight in space had not been considered a complete success, and the very idea of female cosmonauts was not implemented as an indispensable part of the Soviet manned flight program.

Glushko saw an opportunity here. With the support of medical doctors also interested in diversifying the objects of their studies, Glushko approached the government for approval to start the program. It was granted.

Glushko personally took the liberty of inviting me to join a specially selected committee to find appropriate female cosmonauts. After a wide-ranging search—mostly among female scientists, medical doctors, and pilots—a large crowd of contenders underwent complicated and tiresome medical examinations before a much smaller group of them was invited by our commission for interviews. Some of the subjects were well qualified, genuinely interested in the space program, and dedicated to competing for the chance to fly in space. At last, we narrowed the list

2. I do not have to be especially courageous to envisage such a state for many years to come. With an exception, perhaps, in one particular area: the area of space medicine and biology. The very discipline of space medicine could not exist without human beings in space, unlike most of the other scientific disciplines which could conduct their experiments in automated robotic ways.

to three or four of the most appropriate candidates for the forthcoming flights. Glushko invited the commissioners for a small informal meeting. I will never forget his concluding remarks, after each of us had presented our own list of favorite candidates.

He said, "Gentlemen, I urge you now to focus on a most important consideration. While almost any one of these candidates is qualified and deserves the right for spaceflight, think of the millions of Soviet people who will be sitting next to their TV sets and watching every moment of the flight. They will never forgive us if we were to send an insufficiently attractive woman into orbit."

With that, Glushko suggested his own order of priorities. I did not object to his taste, although one had to laugh at the transformation we made from a serious commission into a kind of beauty contest jury. The choice was Svetlana Savitskaya. As a matter of fact, it turned out that she was the only female cosmonaut who flew in space as a result of Glushko's very visible and spectacular initiative.[3]

In general, the Soviet space system created a special breed of people. Heroes of space were selected on a very strict basis of utmost political loyalty. There was a taboo against any undesirable ideological virus that might be launched into orbit.

Of course, no one in the ruling elite was interested in keeping the space environment sterile per se. The secret was much simpler. The heroes of the cosmos, upon return to earth, would be quickly promoted to important political posts. Almost automatically, most of them became members of parliament: deputies of the Supreme Soviet. A few of them even became members of the Central Committee of the CPSU.

The "right stuff" of Soviet cosmonautics was the single, best represented social group in any kind of supreme elected body in the nation.

3. From the point of view of training, Savitskaya was clearly the best qualified, as she had considerable flying experience with almost every kind of modern aircraft. For a number of years she held the records for altitude, speed, and so on among female pilots. What I was unable to approve of ultimately, however, were her political views. Because seating was by alphabetical order, many years later in 1989, I sat next to her at the Congress of People's Deputies. At a most shameful moment, when the huge crowd of peoples' deputies tried several times to humiliate Sakharov by giving a standing ovation to someone who had denounced him in the most blatant terms, I tried not to hear or see this majority. But the person who was seated next to me, and most actively supporting such reactionary sentiments, was female cosmonaut Savitskaya.

Such tight control over the preselection procedure was capable of excluding almost any possibility of future dissent.[4]

After the joint flight of *Apollo-Soyuz* in 1975, the leaders of the military-industrial complex and the space industry somehow sensed that manned flights could bring very important international benefits. As there were no announced joint programs with the United States, the next idea was to start inviting foreign cosmonauts from different countries to fly on our orbital manned flights. What an important political card it could be—an instrument appreciated by the government and Brezhnev himself.

The idea, however, was quickly reduced to a rather cheap political game. Any time Brezhnev prepared to meet one of his counterparts, the head of an allied state, he would take a special souvenir for his political colleague: a ticket to space.

First it started with the leaders of fraternal Socialist countries. Then beyond that circle of "brothers in class," it was extended to a next group of countries considered as having a special relationship with the Soviet Union, such as France. In such a scenario, the leaders of the Soviet space program were always in the limelight.

However, these initiatives didn't yield much. To promote them did not require developing new technology or reaching a breakthrough in another dimension of the space program. Nor did it provide any progress in space science. Moreover, every time the Soviets flew another foreign cosmonaut from a freshly proclaimed space country, a lot of time would be spent teaching the candidate, and even certain groups within the scientific community of the guest country, to invent and to develop appropriate scientific instrumentation for future space flights.

It was not necessarily always rewarding. In many instances we did not encounter any genuine interest in performing space science experiments. However, Soviet authorities pressed us to get something delivered into space, at least to create the image that man is doing something sensible—an orbiting "Potemkin village."

In many cases, independently of how clever the freshman cosmonaut's compatriot scientists were, they had almost zero potential for inventing and preparing special space-qualified experimental devices.

4. In the years of perestroika, a few of my cosmonaut friends became prominent supporters of the radical intelligentsia for the renovation of society. Georgy Grechko, Alexei Eliseev, and Konstantin Feotistov come immediately to mind.

High-level politicians usually pressed on us to find something traditional that represented the culture of the guest country. We at IKI, and our friends in such countries as Mongolia, found ourselves in real trouble. What could be found in the traditional scientific culture of such a country? In an underdeveloped country lost in the heartland of central Asia, the traditional technology probably related to Mongolian shepherds.

While our space establishment's bosses and leaders of the military-industrial complex enjoyed parties with their high-level counterparts, we were expected to produce almost heroic efforts by our fellow scientists and engineers. I felt particularly rewarded that IKI accumulated substantial experience in remote sensing. The program, based on our experience, was probably at its best when we invited the cosmonauts from such vast countries as Mongolia, where the terrestrial resources had been practically unexplored.

One of the trickiest political issues to be resolved at the highest level of the Politburo was who would be the first foreign guest cosmonaut on a Soviet manned spacecraft. It was not so simple to envisage all the political implications of such a first ticket. Of course, it had to be one of the most important of the Socialist brothers. While all of them were equal, someone had to be more equal.

The political intricacy of the selection procedure easily proved to be exciting for leaders of the military-industrial complex. The first ticket to ride was finally given to Czechoslovakia. I believe the main argument was the amount of suffering they had experienced under the domination of the USSR, their "senior" brother. Our friends in Prague were apparently considered the biggest sufferers. The logic was quite understandable from the point of view of the Soviet leadership. There could be no better consolation after the introduction of tanks in Prague than to fly on a Soviet spacecraft. My fellow scientists from Czechoslovakia at that time told a bitter joke: "Why would it be so difficult for Czechoslovakia to build its own satellite? Because it would have to revolve not only around the planet, but also around Big Brother's satellite."

The Poles were given the second ticket. Brezhnev was clearly in a hurry to reaffirm his solidarity with the then Polish party boss Edward Gierek. IKI, again as an important intermediary, was given the honor of receiving the foreign dignitary on the eve of that flight. Gierek himself, with all of his closest associates, visited IKI.

Giving a ride to foreign cosmonauts quickly became an important political hobby in the Kremlin during the seventies. Every time Brezhnev met the leaders of friendly nations, our space community was asked for a new initiative to announce at the summit. The space troika conveyor was already well established, so the answer was automatic: to fly a cosmonaut in space.

The next frontier was to involve capitalist nations in these flights, but only those who had a special relationship with our country. I believe the prospect for the French was not too exciting, especially after a long line of clients had already had their ride in space, including even a former French colony, Vietnam. So the thinking on the French side, according to what I heard from my scientific colleagues in Paris, was to develop in a different dimension. Their government wanted to fly a female cosmonaut. After all, we had heard it would strengthen the political stand of French President Valéry Giscard d'Estaing at the next election. At least it was expected to bring overwhelming support from the best half of the nation.

At the same time, such an initiative would clearly paint the French participation in orbital flights as distinct from the long succession of standard male flights.

Soviet medical experts had provided their French colleagues with long instructions on how to select the right woman from among the candidates, based on rather complicated medical tests. Out of forty contenders to represent France's better half, only a very few were qualified as capable to fly. Among those who were left for further consideration, the selection procedure had to identify those who were scientifically and technologically best prepared. I never saw the final candidates, but soon two appropriate candidates were chosen and preparatory training was started. This created a lot of excitement, especially among the Soviet cosmonauts, and it triggered a strong competition regarding who would fly on the joint team.

Unfortunately, the excitement ended abruptly due to a very trivial mishap. While jumping with a parachute, one of the women broke her leg. The French space authorities suggested appointing a male cosmonaut as a double for the woman. The Soviet space authorities considered such a request and rejected it. They said that the last-minute change of sex would make it impossible to use the same life support systems in flight. Indeed it was a sad story. Who knows, maybe the defeat of Giscard d'Estaing by François Mitterand could have been avoided had there been a better outcome. Eventually we had a couple of rather sophisticated joint flights with French male cosmonauts. Space technology probably did not lose much, but the charm clearly was lost.

One positive development that came out of such manned flights, in general, was the promotion of space activities in participant countries. In Hungary, for example, it gave that country a momentum that eventually led to the establishment of quite an original and ingenious group of scientists and engineers. They later became our indispensable partners in much more complicated space science ventures. IKI established especially good relations with the Hungarian Institute of Physics in Budapest. I have personally enjoyed every minute of this cooperation, in both scientific and human terms.

The proliferation of flights given to foreign cosmonauts created a lot of excitement and political lobbying. More and more countries approached us, asking for the chance to fly. How could we possibly tell them that the tickets were given at the highest level. It was a personal hobby of Brezhnev.

The good old days of free rides in space have been over for a long time. For the last several years, the Soviet—now Russian—space authorities have been acting on a commercial basis: "If you want to ride on the space troika, buy a ticket." In 1987, at a special reception given in the Soviet embassy during Gorbachev's visit to Washington for the summit with President Reagan, I was approached by John Denver. He was absolutely obsessed with the idea of flying on a Soviet spacecraft. His sudden application was not completely impulsive or improvised. He told me a long story about his preparation to fly on an American shuttle mission. He said that originally he had even been scheduled to fly on that tragic *Challenger* mission, but later the schoolteacher Christa McAuliffe had been chosen instead.

After the tragedy, Denver wrote a very moving song in which he paid tribute to the brave heroes of the *Challenger*. However, in regard to spaceflight, it was one thing to be invited by Gorbachev to an important reception in the embassy and another to ask for a ride into orbit. The Soviets demanded $10 million for the ticket.

The last thing I heard, before Denver abandoned this idea, was that a businessman from New York agreed to contribute $5 million for Denver's flight, on the condition that it would be taken only in one direction.

27 / The War of the Worlds

➢

It's remarkable how short the interval was between the launching of the first sputnik and the first probes that were sent to the planets. The very first launch to Mars took place as early as 1960. During the sixties, almost half a dozen spacecraft were sent to Mars, and almost the same number to Venus. Korolev was the originator of most of these launches. The same type of boosters as the ones used to deliver spacecraft to terrestrial orbits were used for distant planets. The only difference could be found in an additional upper stage, which provided a few more kilometers per second of velocity against Earth's gravity.

First attempts to reach Mars and Venus were not successful because they required a different level of sophistication in controlling the spacecraft at farther distances. In addition, the environmental conditions were not well enough known to perform maneuvers such as reentering the atmosphere. In the early sixties, ground-based astronomy was unable to tell the space community what kind of temperatures and pressures could be expected.

The very process of running the space program and the planning of space missions was a rather intricate kind of political interplay—with the interests of different groups frequently in conflict. There was no genuinely simple, democratic approach to making the final decisions about a mission. The space program was run like a sacred rite, looked after by only a few high priests behind closed doors. Korolev, during his time, was able to run everything unilaterally. He was able to push all his projects through channels to get them approved. When Korolev died, Keldysh tried to find a balance between different interest groups.

When I joined the institute, IKI was already engaged in providing scientific instrumentation for four parallel missions to Mars: *Mars 4*, *-5*, *-6*, and *-7*. Two of them were to deliver descent modules to land on the surface of Mars. Four launches took place in 1973, two years prior to the American *Viking*. These *Mars* launches were the spacecraft that developed the terrible on-board virus.

The next attempt to send spacecraft to Mars was originally planned for 1975. That year was the next "astronomical window" for Mars. Keldysh, our colleagues in the Academy of Sciences, and I had a long discussion and, maybe for the first time in our program, had to admit the existence of international interaction, or even de facto coordination. We knew that the Americans were going to send the *Viking* mission in 1975. Some of us raised the issue of whether we should indeed continue to act according to our plans and launch the spacecraft to Mars in 1975 at all.

We came to the philosophical conclusion that since *Viking* was such an expensive, dedicated, and extremely sophisticated spacecraft, we should not try to duplicate the *Viking* mission. Instead, we argued, we should abandon *Mars* and see what would happen to the American program. At the same time our resources should be used for a different target in space. It was a very important decision, which was made at the end of 1973 and early 1974. In fact, it was a real turning point, not only in our program but in the way the decision was reached. I am proud that the original proposal to abandon the launch to Mars in 1975 came from IKI and from me.

Even though there was almost no international cooperation at the time, except the *Apollo-Soyuz* project, the openness of the American program played an important role for us in planning our missions, particularly since the Umoted States would announce all their missions, like the *Viking* and others, in advance.

The first unconscious encounter of the two planning approaches took place even earlier during the most active part of the lunar missions in the late sixties. Definitely, the success of the *Apollo* project and the moon landing helped us abandon the original unsuccessful attempt to send a man to the moon.

To launch a mission and to be the first were the most important motivations for these highly expensive, prestigious missions. If someone had a better chance to be first, it would not be worthwhile to invest so much money in order to be second. We were not engaged in agricultural production, or even commercial space, where everyone would need to have telecommunications satellites independent of whether he was first, second, or tenth.

In a scientific exploratory mission, whoever is first gets the most important information. So the one who is second has to weigh, from the very beginning, how important the step would be. You have to be sure that you would answer critical issues left unanswered after the first mission. It would require additional time to digest and assess the results of the first mission. These were our considerations in late 1973, as we faced the question of whether to fly to Mars in 1975 or not. Detailed knowledge about the nature of the *Viking* project enabled us to see that no second landing would yield scientific information comparable to *Viking*'s.

An asymmetry existed in the very fact that while we knew about the American planning procedure, they did not know about ours. We were approached consistently by our colleagues at NASA asking us to disclose our plans about what we were going to do next with Mars, Venus, and other planets. It was very difficult to persuade the Soviet authorities that we should reciprocate. It was not easy to persuade even Keldysh. The Soviet system had a different culture and mentality.

The very fact that we were doing space science and that after completing each mission the scientific data had to be published openly did not help convince the authorities to disclose all the thinking behind the programs. The authorities had other reasons. Since most of the technical work was done by the aerospace industry, which was part of the military-industrial complex, the authorities were reluctant to put themselves at risk of failing in the eyes of the press and the public. This was not simply imposed by the government or by its agencies. To a very large extent, it was cultivated by the industry itself. In the event of failure, secrecy could provide protection, a kind of escape.[1]

For me as a newcomer to the space community from controlled fusion and plasma physics, my experience with working in the international family of scientists made it very difficult to accept these new rules. I knew it was not an issue of secrecy. It was a kind of game. But secrecy had very long, very deep historical roots in the deformation of the psychology of the people. Those who worked in this area told me that, to some extent, they felt even happier in such an atmosphere of secrecy. Since nothing was announced, there was always the chance to hide the

1. The very history of such purely exploratory scientific missions gave proof of such protection. Official announcements came only after successful launches. However, the space community put an unreasonable burden on itself by not emphasizing the pioneering nature of the science. Failures were regarded as humiliating disasters when, in fact, people were pushing the limits of knowledge all the time.

actual reason for failure, and they could present a final statement not only for general consumption but for government officials, too.[2]

The decision to abandon flights to Mars in 1975 was a turning point in our space program for perhaps another unexpected reason. It was the first time that we felt, even in some small way, we were part of the international community. We waited for what the *Viking* mission would bring, and at the same time we looked to Venus as a more appropriate target for us. We chose Venus deliberately to avoid duplication.

The early launches to Venus in the sixties were done almost in parallel with launches to Mars. By 1975, we were ready to deliver rather sophisticated pieces of hardware to the surface of Venus.[3]

The very first mission sent to Venus in my time was *Venera 9*, and then *-10*, which were landing modules. All precautions were taken to

2. My approach to this question was quite different. I had learned long ago that if you do not hide, then your counterpart won't hide, and both can reach the goal much more quickly. That was quite obvious to me. Science, as an indispensable element of human culture, has been developing this way from the very beginning. In its early days, without all the e-mails, telexes, and telefaxes—even without regular scientific magazines and book publishing—science was made by a few individuals. They wrote remarkable letters, which went into great detail. They requested reactions and answers from their colleagues and their opponents. Science has by its nature always been a collective effort.

3. In the sixties, one of the most successful results we had was to learn the reentry maneuver. The capsule was able to survive, despite its approach to Venus at a very high velocity—about eleven kilometers per second. Once inside its atmosphere, the parachute opened and the capsule landed rather quietly.

In 1967, my colleagues in the Soviet program experienced an extremely exciting moment. One of the early capsules, *Venera 4*, descended into the atmosphere of Venus, every second sending information about the environmental conditions, density, temperature, and pressure in the atmosphere of Venus. Suddenly, the signal was interrupted at an altitude of twenty-five kilometers above Venus's surface. The very fact that the radio signal disappeared at this very moment was thought by optimists to be a result of the impact on the hard surface. Later our American colleagues joked that it was such enormous luck that the very first landing on Venus was able to hit the top of the highest mountain. If such a mountain, twenty-five kilometers high, even existed on the planet it would be three times taller than Mount Everest. One particularly "ingenious" argument used to defend this view was an analogy with the very first German bomb that hit Leningrad at the beginning of WWII. That Nazi bomb killed the only elephant in the Leningrad zoo.

Of course, the *Venera 4* capsule did not land at that moment. It was simply crushed by extreme pressure in the atmosphere of Venus.

protect the lander from overpressure and overheating in the hostile environment of Venus.[4]

Venera 9 and *-10* were the first landers to take into account the implications of these most hostile environmental conditions. They landed a few days apart from each other. Each of them was able to keep room temperature conditions deep inside a protected module for about an hour. But that one hour, calculated in advance, was enough to take images of the landscape around the spacecraft, to take a few measurements of environmental parameters, to code it in digital form, and to send it back to Earth by radio.

Venera 9 and *-10* opened the series of successful *Venera* landings on Venus. Three years later, in 1978, *Venera 11* and *-12* were launched. The original scenario of how to land a quite heavy module on the surface was already well tested, so we invited chemists, geologists, and atmospheric scientists to invent new ideas to make every mission more exciting and more interesting. Each new mission of *Veneras* advanced significantly the sophistication of scientific instrumentation.[5]

At this point in time, our American counterparts had an extremely successful *Viking* project, which on top of important geological and atmospheric findings on Mars, made clear that chances of detecting even the simplest forms of microbic life on Mars were extremely bleak. In any case, the success of the *Viking* mission was overwhelming. We envied the Americans. But at the same time it was a great relief and satisfaction to feel that indeed we had been right to stop our own Martian project. We would have been far behind the Americans in performing on the surface of Mars. Even if the 1975 mission to Mars would have been successful and we had been lucky enough to have a soft landing on the surface, we

4. Both landers of *Venera 9* and *-10*, after reentering the atmosphere of Venus, were able to land happily on the real surface of Venus. The previous designs of the landing modules had not taken into account the actual rate of pressure in the atmosphere of Venus. The atmospheric pressure at the bottom is about a hundred times greater than the terrestrial atmospheric pressure. The temperature is also very high: as high as 500°C.

5. Black-and-white imaging was replaced with color, and an ingenious drilling device was added to pick up a sample of soil for X-ray analysis inside the lander. Both failed at the beginning due to technical mishaps. Later in 1980, *Venera 13* and *-14* finally succeeded in a second attempt.

would not have been able to afford a technical and scientific scenario of *Viking*'s caliber.[6]

So at that point in the space program a very peculiar type of coexistence was established in sharing the duties for deep space exploration between American and Soviet programs.

Viking had been the beginning of the diversification of the American planetary program. The next project was *Voyager*, to fly by several planets. In the meantime, Americans made a flyover of Mercury and Venus with the *Mariner 9* spacecraft. Altogether, *Voyager* and *Mariner 9* made it possible to encounter at least half a dozen celestial bodies, not counting their satellites. It was a wide-range reconnaissance of the Solar System.

Every few years, as director of the Space Research Institute, I had to report, upon the completion of each important mission, to a very large audience in Moscow at the Polytechnic Museum, a kind of counterpart to the Smithsonian Air and Space Museum in Washington, D.C. A lot of space enthusiasts came to ask questions. Every time, while enjoying the new science or experiencing the new excitement of a deeper penetration into the mysteries of Venus, my audience would ask at the conclusion of the meeting "While we are sending spacecraft to Venus—*Venera 8*, *-9*, *-10*, and so on—the Americans fly their spacecraft to visit Mars, Venus, Mercury, Jupiter, Saturn, and so on. Why couldn't we send such missions? This gave me an opportunity to dwell on the major difference between the two countries' resources.

Of course, by comparison the American program was rich with diversity and courage. The most we could do, with what might be given to us by our industry, was to reach Venus and to land there, and to survive for a while in an extremely hot and hostile environment. But to undertake trips to distant planets would require something different. The technology had to be able to work for a very long time, readapting to constantly varying conditions. Such a long survival time would require a different technological culture. It would require a special dedication, which on top of it all would also cost a lot.

However, I thought such an answer would be too boring, so I answered the question in the following way:

"You know, we have a silent, gentleman's agreement to share responsibilities in space. While the Americans are doing wide-range recon-

6. We had to work in a different "weight category": *Venera*-type missions were worth 100 million rubles versus the $800 million *Viking*.

naissance in the Solar System, we are carrying on an intensive study of Venus, just as if that planet were declared the planetary test range of our space science."

But the next time, reporting on the newer *Venera* landing, I had to confront the same questions. Once I was literally bewildered, and without much thinking, immediately answered the question of why Americans do such and such:

"Because they are sons of a bitch," I replied.

For several minutes, the audience applauded. It was very clear that they were applauding the Americans and their space program, which had captured the imagination of the Soviets despite the attempts of our official propaganda to undermine the achievements of these "sons of a bitch."

With such a great variety of approaches, the flexibility of the American program created in us a kind of inferiority complex. In large part we were bound to much closer targets in the Solar System because of our technology. To fly for a long period of time would require extremely redundant and survivable payloads. Of course, we envied NASA's capability to produce such long-lived spacecraft. One would have to forget about the distant planets if one's spacecraft could not survive in space for more than at least a few years. We had a claustrophobic feeling that we were serving a lifelong sentence on Venus. Once retreating from Mars, we deliberately and consciously avoided competition with NASA, and at the same time freed up our resources to start a long and rather successful, if not eventually boring, series of flights to Venus.

Toward the end of the 1970s, the Soviet space science community was divided into different schools of thought. One group, predominantly those who were already enjoying flights to Venus, was perfectly happy and counted on continued exploration of the hot planet—one flight after another. Venus could keep space scientists busy almost forever. Another very strong group was pushing for a most expensive project, aimed at bringing back a piece of Martian soil so final experiments on soil analysis could be conducted in the comfort of a terrestrial laboratory.

The driving force behind these sample lovers was the strongman of Soviet earth sciences, academician Aleksandr Vinogradov. Approaching his eightieth birthday, he was the academy's vice president, responsible for all of the national earth sciences. He also directed the Institute of Geochemistry, originally established by famous ecologist academician Vladimir Vernadski. Vinogradov did so with support from the military-industrial complex and the party elite.

I could never figure out how this short, almost miniature, old man with white brows and a soft-spoken manner could influence the decision-making process up to the level of Marshal Ustinov himself. In principle, I was in full support of the concept of a mission to bring back a Martian sample. But I did not consider this mission realistic within the short time frame, which was in the interests of the aging Vinogradov. In my view, to send a rocket to Mars—capable of flying there, landing on the surface, and eventually coming back to Earth—space technology would need to progress for at least fifteen or twenty years more.

Keldysh rather passively watched the development of the debate without expressing his own view on the project. I even tried to provoke him into taking a position in the forthcoming war between "Martians" and "Venusians." I had an instinct that the reason for Keldysh's indifference was a kind of old antipathy toward Vinogradov. The geologist had been Keldysh's own choice for the vice presidency of the academy. However, they did not develop an intimate cooperation. While Korolev and Keldysh took responsibility for preparing the lunar missions, Vinogradov somehow secured and monopolized the lunar samples. He was smart enough to understand that this extraterrestrial material, eventually delivered back to the earth by Soviet robotic spacecraft, in fact was an extremely powerful political instrument.

Soon, Vinogradov found an international dimension for such a public relations campaign. Every foreign institution willing to participate in the lunar sample analysis would have to undergo an almost religious rite, with the indispensable glorification of academician Vinogradov himself.

Vinogradov looked rather impressive: a little man capable of delivering a persuasive scientific sermon. My first reaction when I met him at the beginning of my space career was: "This man is too academic. He must be very far from the political scene." However, very soon I learned that a few years earlier Vinogradov had implanted one of his close associates, Valery Barsukov, as an important party apparatchik—politcommisar of Soviet earth sciences—into the very stronghold of the Communist party: its Central Committee. Vinogradov proved himself, in fact, to be a master politician.

Of course, from the very beginning the chief designers in the space industry, thought that it would be too difficult to complement Vinogradov's collection with Martian samples, at least in the twentieth century. However, they liked to be courted by the old academician, an influential courtier in higher circles. In reality not one of them thought that such a project would ever be implemented.

However, they did not take into account the inner logic of irresponsibility, as it shaped the pace of events in the overall stagnating atmosphere of the late seventies. Marshal Ustinov, after several phone conversations with Vinogradov and visits to the "holy shrine of lunar samples," was converted into an ardent supporter of the idea. After all, he himself had not heard from the rocket professionals in the space industry that this project was unrealistic. Besides, nothing would please his personal ambition more than leaving something essential for eternity. What could be more impressive than Martian soil?

Knowing the mentality of space industrialists and politicians, I believe there was one more very important, even if hidden, argument. The authorities were bored with the continuous lobbying we scientists were conducting for one more scientific mission, one more instrument. Irritated by such begging, they had a secret dream to pick one proposal, out of a huge number of scientific ideas, which could be implemented as the ultimate science, which would discover and unveil all of nature's secrets.

I think academician Vinogradov, in his campaigning for the Martian sample return mission, subconsciously hinted that after this there would be no more mysteries about the Solar System and its origin. The Martian sample would be, for space scientists, a magic philosopher's stone.

How congenial it all sounded to such a breed of Bolshevik as Marshal Ustinov: the ultimate space science as an accompaniment to the supreme instrument of knowledge—scientific communism.

One day, in the early winter of 1976, a huge crowd of chief designers and the directors of different mailboxes, mingled with a few scientists, were invited to the principal design bureau for planetary and lunar missions—the same institution where Babakin had successfully established the first robotic projects to the moon and the planets. It was a Saturday morning, the beloved time for Ustinov to organize such gatherings. I was almost sure that someone among the space industry's professionals would not evade the truth of the matter any longer and would finally confess that we did not have the technology to perform such a sophisticated mission.

No one, unfortunately, was brave enough to acknowledge the truth to Marshal Ustinov, preferring to remain silent over the deficiencies of such a premature project. The only precaution the professionals assumed was great timidity about taking the project's leadership role for themselves. Eventually, Marshal Ustinov picked a rather young and overambitious engineer, who was not even a chief designer at that time, and nominated him as technical director of the project.

In the suggested scenario, IKI would have no specific responsibility and would never be at any risk, except with respect to the moral effect of being present at such a discussion of the emperor's new clothes.

I was not an expert in space transportation, and my judgment on whether such a scenario would be technically feasible was not going to be decisive in the discussion. However, witnessing the professionals' reluctance to be direct with Ustinov, I tried to attack the project from a different point of view: "The international scientific community attributes great importance to the risk of bringing back to the earth alien living species, microorganisms. The sample brought back might contain a virus from Mars. This could have unpredictable consequences. To eliminate such a risk, scientists suggest different scenarios, like the establishment of an intermediary receiving station for the sample, in orbit around our own planet, with a subsequent period of quarantine."

In fact, there was no exaggeration in such a proposal. For a number of years, discussions of this type were organized by a special international scientific commission on planetary quarantine and sterilization.[7]

The old academician immediately sensed that my indirect counterarguments could finally torpedo his project, and he immediately entered into the debate. "Academician Sagdeev raises a very important question. We Soviet scientists, at the forefront of planetary exploration, have to pay very serious attention to these issues of back-and-forth contamination. I myself have a specific proposal. First, we should eliminate an expensive intermediary station in terrestrial orbit. Second, to ensure that we are not bringing to our own planet a deadly Martian virus, the sample should be treated with the most efficient chemical substance, like poisonous gas, to kill all potential microorganisms."

I tried to interject "But, esteemed Aleksandr Pavlovich, the life

7. The opposite side of the coin was the danger of contaminating extraterrestrial bodies—in this case the planet Mars, with Earth-borne viruses unintentionally delivered to the surface of the planet. The international scientific community had been pressing on space-faring nations to ensure that no such contamination take place.

scientists and exobiologists would be waiting impatiently for this Martian sample, to determine whether or not there is life on Mars. We don't need war on Mars."

Vinogradov coolly said, "So what, they will have a chance to look for dead bodies, the corpses of this hypothetical species."

I don't know whether Marshal Ustinov was particularly impressed with this last argument of Vinogradov's or, as Minister of Defense, he was fascinated with the prospect of launching and winning a small controllable chemical war against Martian viruses—that is, if they did indeed exist.

The end of the story was very typical of Ustinov. He asked questions: "How many years would the space industry need to implement such a project?" Rather cautious and vague answers quoted something around ten years. In his concluding speech Ustinov said that it was absolutely unacceptable for the Central Committee and its Politburo. "I will give you three years," he said firmly.

Ustinov knew his subordinates—the only argument they appreciated was sheer pressure.

Afterward, everyone was invited to a tea party, at which the very first traditional toast was raised for dear Dmitri Ustinov, the guardian angel of the Soviet space program. The second toast was suggested for the health of Mstislav Keldysh, the chief theorist of the space program. I had the impression that Keldysh was unusually sad at this party. He seemed detached from the reality of life or the absurdity of it. He certainly kept some distance from the decision that had just been made. Even the Big Hammer, the minister of the space industry, who clearly was a realist, did not raise any objection—giving up the power to Ustinov.

Within a few weeks, the final official document was issued by the Central Committee and the Council of Ministers, decreeing the launch of a sample return mission to Mars. It was delivered by secret mail. I asked my friends and supporters in the apparatus of the military-industrial complex, the friends of IKI and space science, "Don't you think this is completely ridiculous? The bosses must be insane to talk about such an adventuristic mission."

With smiles characteristic of the veterans of bureaucratic life, my interlocutors responded, "Don't worry. The military-industrial complex is polluted with such papers. Everyone here knows the secret. Only sixteen percent of the plans formally approved by the Commission on Military-Industrial Issues ever eventually materialize. And as a Tatar, you should follow the Oriental wisdom of Khoja Nasruddin, the legendary hero of medieval fairy tales, who courageously promised the shah to teach his donkey to speak the human language. After being asked by shocked friends how he could make such a promise, he responded, 'I am sure one of them will die first, the donkey or the shah.'"

I don't want to draw a parallel. However, as sad as it might sound, esteemed academician Vinogradov did not live long after that famous session, and the glorious marshal, moving more and more away from civilian space, had probably completely forgotten about his promise to support that ill-fated mission.[8]

8. In 1979, the year scheduled for the launch, Project 5M (the code name used for the Martian sample return) was officially abandoned. I think it would be very interesting to follow the files of the Commission on Military-Industrial Issues and the Ministry of General Machine Building Industry to find out what kind of final official excuse was given for abandoning the fulfillment of this important governmental decree.

28 / The Hostage to Politics on Earth

➢

In November 1975, a few days after that famous successful landing on Venus, in the wake of the triumph's ongoing television and press coverage, I was visited by Roald Kremnev, a space engineer from Lavochkin Design Bureau, our principal contractor in the industry. With a conspiratorial confidence established between two namesakes, Kremnev whispered to me, "I have been to the Defense Department of the Central Committee and heard that your name is on the draft list for the Lenin Prize."

Nothing could flatter a scientist's vanity more than such a decoration, considered the highest scientific award in the country. In addition to the indisputable honor it would have, it also brought a lot of money—about 1,500 rubles, equivalent to two months' salary. That would be roughly sufficient to invite a crowd of friends and colleagues for the traditional—and absolutely unavoidable—banquet.

Bestowal of the Lenin Prize for space achievement during the pre-glasnost epoch was not publicized in the press. I expected instead to hear of it instead in the form of a congratulatory phone call from the top. Indeed, a day or two later I received a telephone call. The secretary of the then–acting president of the Academy, academician Vladimir Kotelnikov, called me on the telephone. In a rather unusually solemn voice, he requested my visit to his office.

"When?" I asked.

"Immediately," was the answer.

I jumped into the front seat of my official black Volga, trying to calm my nerves in anticipation of the forthcoming honor. The corridors of the academy's headquarters, the presidium, were strangely overcrowded. There were a number of lines of academicians waiting for their turn to enter the high offices. I quickly figured out that such a "conveyer" couldn't be for the mass production of Lenin laureates. Something else must be going on here; I understood immediately that it must be related to Sakharov bashing.

The name of Andrei Sakharov was a firmly established irritant for the authorities since he had broken with the nuclear establishment and written his eloquent manifesto, "Reflections on Progress, Peaceful Coexistence, and Intellectual Freedom," in 1968. With that he became the country's most prominent dissident, whose human rights activities created a growing concern for the Brezhnev government. I witnessed the evolution of Sakharov's civil consciousness with a blend of genuine admiration and fear. This was a typical attitude among us scientists, who praised Sakharov only in the narrow circles of our internal exile—at the kitchen table. Most of us were simply not brave enough to publicly defend Sakharov or his ideas. It would have meant the risk of losing our petty privileges.

Even while completely absorbed by our space odyssey to Venus, I couldn't help but notice, in the autumn of 1975, that Sakharov was awarded the Nobel Peace Prize. The announcement of it almost coincided with the landing of the *Venera 9* and *-10* modules on the surface of Venus. It was with great displeasure—even outrage—that official circles in Moscow learned about the decision of the Nobel Prize committee. They wanted immediate retaliation, delivered by Sakharov's own scientific colleagues—the full members of the Academy of Sciences. That is the reason I had been called so urgently by Kotelnikov.

Without wasting much time, Kotelnikov handed me a sheet with a draft letter of condemnation, already signed by a number of academicians, who had had the "luck" to hurry to the presidium before me. Written in a most blatant style, I understood that the draft had not originated in the Academy of Sciences; it must have been written instead by the Party ideologues at the Central Committee. When I hinted this to Kotelnikov, he didn't deny it.

I was in a very delicate position, desperately trying to find an appropriate way to express my position. I told Kotelnikov that I could not sign a letter written by someone else. He coolly responded: "Try to draft yourself another version, if you can. I am sure nothing will come of it; you will have to come here once more to sign this very text."

I went back to my office with the naive hope that I would not be bothered anymore. Alas, the academy clearly had instructions from the Central Committee to demonstrate the unanimous anti-Sakharov feeling among the leading scientists of the country.

A few hours later, Kotelnikov phoned me again and asked whether I had drafted my own version. "If not, you have to come immediately and join us."

That dreaded prospect prompted me to squeeze a few sentences of a compromise denunciation out of my own conscience. After the text was typed I instructed my secretary to destroy the original and to keep the story confidential.

When I reached the presidium again, Kotelnikov was obviously displeased that I had come with a separate statement. "Don't you understand you are putting yourself in a special position by demanding a separate publication?"

As an antidote to his comments, I responded, "Vladimir Alexandrovich, I *am* in a special position. I was a student of Andrei Dmitryevich, during my diploma work."

I spent the next day at Keldysh's Institute of Applied Mathematics, conferring with a large group of space scientists on our future plans. At the end of the meeting Keldysh asked me to stay for a few minutes for a tête-à-tête conversation.

"I heard you did not sign that letter on Sakharov. I think you are making a mistake and you might be sorry for it in the future."

By that time *Izvestiya*, the governments's official newspaper, had already published the letter of condemnation. It was signed by seventy-three members of the academy, including both Kotelnikov and Keldysh.

The rumors spread quickly that only a few academicians had refused to sign. Among them were Peter Kapitsa, Yuli Khariton, and Yakov Zeldovich. Apparently the last two had insisted on being excused as former colleagues of Sakharov. My independent letter was never published. Apparently the authorities considered it unacceptable. They clearly didn't like my attitude that "Sakharov must be reproved at most for abandoning his active scientific duties for which," as I stated in that letter, "he undoubtedly could win a Nobel Prize for science."

Needless to say, I was perfectly happy that this letter, a compromise with my conscience, was never published.

Retaliation for not marching in step, however, was not long in coming. The first indication came with a phone call from Kremnev, my namesake, who said: "Roald, I heard bad news at the Central Committee. Ivan Dmitrievich Serbin, said 'Sagdeev will never get a Lenin Prize and he knows why.'"

Not long after that, I was asked to visit the Central Committee. Serbin, the same politcommisar of the Soviet defense establishment familiar to me since the Geneva conference "Atoms for Peace" in 1958, was attentively studying the expression on my face. "Do you enjoy being director of an institute or not?"

I am sure he was surprised by my answer. "I haven't yet made up my mind on that issue," I said.

"Well, we have checked your claim of being a direct pupil of Sakharov. It is unjustified."

I immediately understood that in the eyes of the government I had no excuse for not signing the letter, and that they were issuing a warning.

In the following weeks, I felt that a peculiar polarization had suddenly developed among the people I usually met and talked with. My counterparts, the bosses of different space industrial enterprises, behaved as if I had contracted a contagious disease. Most of my scientific colleagues and junior collaborators, on the other hand, clearly expressed moral support, even if it was only whispered.

The Lenin Prize was gone. And I better deserved to be part of the group Lenin used to call the "rotten intelligentsia." The list of punishments for this offense included more than losing the prize. They also took away the institute's *kremlevka* telephone that linked me to the *nomenklatura* and the space industry bosses. In a telephone call I made to "Uncle Mitya"—Comrade Ustinov—in his office in the Politburo of the Central Committee, I tried to explain to him that cutting my *kremlevka* phone line to punish me would hurt the institute more; it would leave the whole institution without such a link. He couldn't hide his irritation and within days my *kremlevka* line was gone. At the same time no one actually tried to kick me out of the director's office. The authorities made it clear that I was not yet irreversibly labeled a dissident, but that I had been "drafted" into the "punishment battalion."

It was that same time, in late November 1975, that the Academy of Sciences, established by Peter the Great, celebrated its 250th anniversary. This anniversary was inaugurated by Leonid Brezhnev. Entering the Kremlin Palace, lent by the government to the academy on that occasion, I met Sakharov, who looked at me in such a significant way that it was obvious to me that he knew about my refusal to join the chorus of condemnation. Sakharov was in a conciliatory mood, even if deep inside he still felt bitter about the recent campaign of character assassination.

After all, the occasion provided a momentary opportunity to detach oneself from the unpleasant reality of Brezhnev's Russia.

The jolly, jubilant atmosphere of the meeting, however, was short-lived. After Brezhnev had finished his deliberations, a rumbling in the audience made us turn our heads. It was Sakharov, who was trying to approach Brezhnev, probably with a letter. However, the row of vigilant plainclothes security officers prevented the scientist from stepping up to the podium. Sakharov tried to break in, but was unable to overwhelm the physically dexterous young men.

Being sucked into the turbulent black hole of the space program, with its hot debates about projects and launches, I didn't see much of Sakharov during his last few years in Moscow, before he was exiled to Gorky. I was trying to rescue space science and the missions to the outer planets, while he was doing the genuinely noble work of trying to save the honor and consciences of us, his compatriots.

In dealing with the space establishment and government circles I quickly discovered I was a defective defender of space science. Even the hint of my sympathies for political dissidents on the one hand, and the independence and disobedience in my everyday contacts with authorities on the other, resulted in the loss of support for the aspirations of the Space Research Institute. The clear-cut "inferiority" of its director in the environment of the military-industrial complex was aggravated by my longstanding reluctance to formally join the Communist Party. In that epoch of "developed socialism," as Brezhnev used to call it, directors of such institutes were rarely recruited from among nonparty members. I was obviously a "white crow" who deliberately avoided the standard test of loyalty.

I had resisted. But such resistance, to some extent, was in the hearts and attitudes of many, especially among the intellectuals. This was in contrast to the behavior of quite a large group of people who were active seekers of party membership. Our society had produced a whole new breed of people—people who from an early age knew that their future promotion in the world depended on both their public standing and their political activity.

From my very first day in the office of director, I became an attractive target for party bosses who wanted to convert and "baptize" a new party member. I joked that I was too old to go through all those things. However, the truth was that the party organization played an important role in the everyday life of any enterprise or institution. I was asking myself incessantly whether I had a moral right to keep my chair as director if I was unwilling to join the party. My reluctance was damaging the interests of the institute. Should I sacrifice the good name I hoped I had in order to promote the institute, its space science, and the ideas of my colleagues and friends? It was with those arguments that I finally persuaded myself in 1976 that I should tranquilize my conscience in order to justify joining the Communist Party.

Since the nominal motto of the party was the "dictatorship of the proletarians," only a small number of vacant chairs could be filled with representatives of white collar groups. My scalp was considered quite an important trophy, and the party secretary of the district made a number of trips to meet me, personally.

To become a member of the Party, an applicant had to supply three recommendations from old established members. In my case, one of those party veterans was academician Aleksandr Prokhorov, Nobel Prize winner for the discovery of lasers. At that time he was the head of the Department of General Physics and Astronomy in the Academy of Sciences. I admired him not only for his scientific talent, but for his civil courage. Not long before that time, he had had to face a grave incident as editor in chief of our Soviet encyclopedia. The party censors had not let him even mention the name of Sakharov under the letter *S* in "Who's Who in Science." Prokhorov took all the risk of acknowledging the very existence of such a scientist named Sakharov. Prokhorov had to go up to the level of chief party ideologue Mikhail Suslov.

I cannot indulge or forgive myself for the fact that I succumbed and joined the party, the monstrous superstructure built to promote the stability of the ruling elite. But to a large extent, I was able to give myself an honest accounting. I was not a man who could or would steadily climb to the top of the administrative hierarchy, even the scientific one. I always felt a deep-seated personality split being a scientist and a boss at the same time. Perhaps the pressure had been too strong and I had gotten tired of trying to carry on a circle defense for the cause of the institute and science.

The party system, with all its seemingly impeccable stability, already carried the virus of decline and self-destruction. Micromanagement of everything doomed the system: micromanagement whereby every step, every move on the lowest level had to be sanctioned by the party hierarchy. That included the precise number of man-hours space researchers of my institute had to spend at the harvest, complementing the efforts of the kolkhozniks, the peasants from the collective farms near Moscow. Official party jargon of those years called it our "patriotic duty on the food front." The list of such extracurricular activities or obligatory community service was controlled by party bosses. It was almost infinite and it continued to grow. Eventually, the biggest time-consumer for the scientific intelligentsia in Moscow was the huge network of monstrous warehouses—the "vegetable bases." Many of my colleagues acquired tremendous qualifications in packing potatoes and carrots under the direct supervision of the highest party officials.

The number one man in Moscow—secretary of the city party committee and member of the Politburo of the Central Committee, Comrade Viktor Grishin—had a special obsession with making sure that no one escaped this honorable duty. From time to time, I got telephone calls from the district party boss, informing us—in a state of panic—that there would be a forthcoming on-site inspection by Grishin, during which he would demand a fresh contingency of IKI scientists to double as "potato experts" and go to the vegetable warehouse assigned to the Space Research Institute. There were stories about how prideful scientists would leave their professional business cards in every pack of potatoes, testifying to the fact that this particular job had been done by the winner of a Lenin or State Prize, and of a Ph.D. in physics.

Micromanagement extended even to the level of family affairs. Application forms or recommendations for promotion for a foreign trip had to confirm that Comrade So-and-so was an exemplary family man. Otherwise, not infrequently they would use the standard formula: "The reasons for divorce were studied by the party committee." Party micromanagers were ready to resolve all sorts of issues. The following popular joke explains it best:

The party secretary: "Comrade Ivanov, I have a complaint that you are avoiding your marital duties."

Ivanov is clearly embarrassed: "Esteemed Comrade Secretary, I have to confess to you, I am impotent."

The grave voice of the secretary: "Comrade Ivanov, first of all you are Communist, and only then may you be impotent or whatever else."

The everyday routine control and intervention of the party, even in the tiniest aspects of the country's life, created an extremely rigid and inflexible system. Unable to support incentives for economic efficiency, it led step by step to the deterioration in the supply of basic goods and food—and finally to the discreditation of the party in the eyes of the population, the ultimate consumer of potatoes.

Established order showed the cracks of imperial fatigue, while the regime was more and more involved in celebration and self-glorification. Almost every new appearance of Brezhnev himself on the podium of the Lenin mausoleum revealed a newly acquired award or decoration on his chest. That's how he eventually became fivefold hero of the Soviet Union and of socialist labor combined. Some of these medals required the rewriting of World War II history to depict Brezhnev as the greatest strategist on the Eastern front.

It was not long before the best brains in the Central Committee had almost exhausted themselves. They were unable, at least for a few brief moments, to invent any other honors for Brezhnev.

"Eureka!" one of them probably exclaimed, "Brezhnev has been a member of the Party for just over fifty years." A special commemoration medal was introduced: "Fifty Years in the CPSU." Another huge celebration was held, this time followed by a chain reaction of mass recognition for everyone with a similar tenure in the party. My father in retirement in Kazan, hit the lucky lotto. Even the cynics who considered such a medal a chunk of metal had to admit that the accompanying letter was priceless: "The regional party committee of the Tatar Autonomous Republic sanctions the name of party Comrade Zinnur Sagdeev to be given access to a special food store in Kazan."

On one of my visits, I couldn't resist making a joke about coming from Moscow to congratulate him, but I am afraid he didn't like my humor: "Father, now you have gotten all you fought for in the revolution."

My words did not irritate my mother, however. She took everything in a much more pragmatic way and told me, "Son, now you won't have to bring meat from Moscow anymore. We are entitled to receive a few more kilos every month."

The nationwide celebrations, campaigns, and mass distribution of honors did not touch Sakharov. Quite the opposite. His open protest against the invasion of Afghanistan resulted in Brezhnev's decision to strip Sakharov of all three medals of Hero of Socialist Labor and the Lenin Prize. The complementary political exile in Gorky immediately meant a shortage of foodstuffs, if not full-scale hunger. We Russians knew well enough the geographic peculiarities of the food supply.

Sakharov had not been in Gorky for too long before he declared a hunger strike in the autumn of 1981. All of us in Moscow's academy circles became extremely concerned with Sakharov's intentions. The story was very simple and clear: His stepson Alexei Semenov, who had emigrated to the United States, wanted to reunite with his fiancée, Lisa Alexieva. Soviet authorities quickly figured out they had been given a lucky break to retaliate against Sakharov and his wife, Elena Bonner, and they refused to give Lisa an exit visa.

Exhausting all civilized legal measures at his disposition, Sakharov decided to undertake extreme measures and issued in advance his warning of a hunger strike. However, the party bosses ignored it. In our kitchen-style think tanks in Moscow, we made an assessment. The party strategists were counting on the understanding, and maybe even the support, of international public opinion, based on the fact that such a hunger strike might be considered rather ridiculous. Pressure was exerted on members of the Sakharov family to make sure they would stop him instead of allegedly forcing him to launch a hunger strike.

We were extremely alarmed with the attitude of the authorities, who obviously underestimated the seriousness of Sakharov's intentions and the graveness of the potential consequences. Three of us, my old friend Velikhov and Vitaly Goldanski, a prominent chemist, drafted a letter addressed to the authorities explaining the worst-case scenario of what might happen and advising the government to let Alexieva go. The letter was to be delivered by Velikhov to the most appropriate boss he could reach. By that time, Velikhov was already quite prominent not only in the academy (as vice president), but also in military-industrial circles as one of the leading contractors for high-powered lasers.

I had absolutely no idea how the inner thinking of Brezhnev's cabinet was evolving, but the state of Sakharov's health was deteriorating rapidly. In early December 1981, I asked for an audience with the president of the academy, Anatoly Alexandrov, the famous old fox. I thought it would be absolutely unbelievable if the Academy of Sciences, the colleagues of Sakharov's own alma mater, silently witnessed the approach of such a tragic outcome.

While waiting for the appointment, I discovered that another physicist, academician Arkady Migdal,[1] was also waiting to see Alexandrov.

Probably each of us at the beginning was quite hesitant to discuss his mission, but then we discovered that we had been brought to the presidium by the same cause, the issue of Sakharov's hunger strike. Thus, we built a united argument for Alexandrov. In fact, Alexandrov himself was seriously concerned with the situation, and he tried independently to find several alternatives for action. "It's absolutely futile to talk to Brezhnev," said Alexandrov. "I think Brezhnev is almost incapable of comprehending anything."

"Why not go to see Andropov?" I suggested. "He is probably most influential. At least he can better understand the tremendous negative impact that a disaster with Sakharov would undoubtedly create internationally."

I never spoke with Alexandrov again on the issue, so I don't know which of the actions finally worked. I'm sure there were multiple initiatives coming from many quarters. In any case, they eventually reached the goal and the government retreated. The recollection of these episodes now makes sense after Sakharov's own description of the hunger strike story in his memoirs. It is clear he was extremely conscious and sensitive to the reaction among his scientific colleagues to his hunger strike. However, he bitterly complained that he did not see any attempt among his fellow scientists to help him even tacitly "behind the scenes."

After Brezhnev was gone and Yuri Andropov inherited the power as general secretary of the CPSU, there was hope everywhere, if not for a complete reshaping of the policy at the top, then at least for more common sense and rationality. In the spring of 1983, the first spring without Brezhnev, I got a telephone call from an official telling me that the Central Committee was planning to publish an article about Sakharov.[2]

"So, what's wrong with him?" I asked.

Apparently, just a few weeks earlier, the Western press had published an open-letter exchange between Sidney Drell—a distinguished American physicist and arms controller—and Sakharov. Sakharov's remarks

1. Migdal was from an older generation of pupils of Landau and Kapitsa.
2. Sakharov was still in Gorky.

were considered support for the new American MX-rocket program. It immediately infuriated Soviet leaders. In addition, they thought that now they could argue that Sakharov had betrayed the interests of Soviet national security. The smart strategists in the Central Committee this time decided that instead of collecting the signatures of a huge crowd of academicians, they would avoid the impression of an orchestrated mass condemnation and invite only a small group of celebrities. ("We don't want overkill!")

A draft article had already been prepared with the title "When Honor and Conscience Are Lost."

When I got the call I said that I was planning to leave Moscow on vacation, and I would have no time to read all the materials to learn more about the article.

"Oh, we don't want you to spend much time at all on it. You simply have to trust us and agree to lend your signature. It is already such great luck that I have caught you before your departure."

My reply was: "I suggest you consider you had bad luck, and that I have already left Moscow on vacation."

In retrospect, I believe that my suggestion on the telephone was taken. Later, I never heard any comments about my refusal to sign this letter.

On the same occasion, my friend Goldanski was invited to the Central Committee and given the chance to read the draft article, but he refused to sign it. As he recalled later, he was treated as persona non grata after that. A couple of times he was refused permission to travel abroad, perhaps for this very reason.

The article was published in *Izvestiya*. It was signed by four academicians, three of them whom I had never respected. After all, even their scientific achievements had not been impressive. However, the fourth name was a real shock to me: Nobel Prize winner Prokhorov. It is still a mystery to me. I cannot explain how on earth he was forced or intimidated to do so. The whole article, written in the most blackened expressions of denunciation, sounded like Brezhnev's postmortem revenge.

The former general secretary was awarded, or rather awarded himself, with every imaginable or even unimaginable medal and title. However, there was one award he did not get: the Nobel Peace Prize. Clearly it was intolerable for him. Henry Kissinger had received it. Menachem

Begin had received it. Even Anwar Sadat, Brezhnev's former client had been given a Nobel Prize. But here, in his own home country, in his estate, where he was in control of everything, where it was in his power to strip anyone of any medals and decorations and to send them from Moscow to, for example, Gorky, there was one person decorated with the Nobel Peace Prize—a person who had created so much pain and humiliation for the omnipotent general secretary Brezhnev.

The anonymous black humorists invented a story. Brezhnev, desperately seeking new awards and glorification, approaches Lenin in his tomb in the mausoleum. Somehow he persuades Lenin to take a short walk outside, to familiarize himself with the achievements of Brezhnev's rule. Upon his return, the founder of the Soviet state discovers his place is occupied by Brezhnev. All the attempts to make him leave the coffin are in vain. At last someone tries a trick. "Comrade Brezhnev, foreign radios report you have been awarded the Nobel Prize."

I believe that, deep inside, Brezhnev was never able to forgive Sakharov his Nobel Peace Prize. And that was the real motivation for the cruel campaign of repression against the scientist.

29 / Big Oleg

➢

The nuclear arms race of ICBMs and SLBMs[1] escalated during the Cold War confrontation of the late seventies and early eighties. It brought with it another, perhaps more dangerous, dimension: the massive buildup of medium-range nuclear rockets in Europe. In the Soviet arsenal, the deadliest of rockets were known in the West by the name SS-20. They were carried by trucks, which made them mobile, thus practically invulnerable. They were counterbalanced on the Western side by Pershings.

Future historians may uncover who was the first to provoke this Euro-strategic arms race, but at that time the Kremlin elite was paranoid about the technical capability attributed to Pershing rockets. According to leaks in the press, allegedly the accuracy of these rockets was enough to "kill the Soviet leaders in their bunkers." Whether the apocalyptic scenario was right, nobody knew for sure.

Though perhaps we might smile now, at that time the Politburo took the possibility seriously. No one could be trusted more than Konstantin Chernenko (a personal confidant of Brezhnev since the forties) to supervise the colossal subterranean project of digging a system of bunkers and tunnels under the Kremlin and the Central Committee. For this, Chernenko was awarded a secret Lenin Prize. The grateful old-timers in the Kremlin later elected him general secretary during the quick series of successions caused by the natural extinction of party dinosaurs Brezhnev and Andropov. Even so, Chernenko himself didn't stay more than a year.

At my end of the party hierarchy, the prize I got for stooping so low (and joining the party) was the eventual establishment of reasonably good relationships with our local party masters. It helped the Space Research Institute to dig its own, safe, ecological niche in an otherwise dangerous and hostile environment. According to the code of such relationships,

1. Submarine-launched ballistic missiles.

we did little favors for each other. Party leaders of my district were given a chance to bring their important guests on sight-seeing visits to the institute. We were available to impress the distinguished visitors with stories and demonstrations about the space odysseys of science and, at the conclusion, to invite them for a formal tea party with an abundance of toasts to strengthen the ties of friendship.

In return, the institute established an uneasy coexistence with the external administrative world that led, finally, to the stabilization of the institute's financial and logistic base. IKI was able to renegotiate the heavy quotas of forced labor conscription imposed by the district party authorities at harvesttime. It also brought material dividends to the staff, with special deliveries of goods that were in shortage, in the form of extended quotas granted by the district party committee.

The system might have been doomed, degenerated by the network of clans and mafias that infiltrated the party and government structure; but within the system's lines, an institute of space research was capable of laying the major groundwork for some very ambitious space projects. Our weak point was the indifference of old-timers, like President Alexandrov, and the fallout from the space wars against the Big Hammer, the ebullient czar of the space industry.

I never could have imagined in the seventies that I would feel nostalgic about "Big Hammer" Afanasiev.

Afanasiev's successor, Oleg Baklanov, was known in the space establishment as a man of a completely different manner. He was courteous and low-key. If not for his slightly southern Ukrainian accent, one might have thought that he was part of the Moscow intelligentsia. I witnessed his appearance on the political horizon of the Soviet space industry in the early seventies. He moved from the position of "engineer" to "administrator," as the director of a huge enterprise in the eastern Ukrainian industrial center, Kharkov. In that capacity, he was one of the principal subcontractors for on-board electronics and control devices. His promotion to Moscow's headquarters of the Ministry of General Machine Building marked his acceptance into the inner circle of the military-industrial *nomenklatura*.

I think the supervisors from the Defense Department of the Central Committee had already, by that time, singled him out as one of the potential cadre for higher promotion. He was much easier to deal with, compared to the Big Hammer, who was often rough and uncompromising even in comparison to the politcommissars in the Central Committee.

For a number of years Baklanov modestly occupied the third-ranking chair in the ministry, waiting for his stellar hour. The more power acquired by "Uncle Mitya" Ustinov, the less time that remained for the marked Big Hammer. Party officials from the Central Committee's Defense Department nurtured Baklanov, as coaches prepare a new rising football star before he enters the game to replace an aging celebrity.

The final removal of the Big Hammer was a great shock for the space empire. All of us recognized the unchallenged czar Afanasiev. But for his main rival, Ustinov, it was one of the last showdowns. The omnipotent Marshal Ustinov's health was quickly failing and his political influence in the Politburo was already diminished. Andropov, in his short career as general secretary of the CPSU (after Brezhnev), had enough time to demonstrate that he was by no means under the spell of the head of the military-industrial complex.

The story of the downing of the Korean airliner over Sakhalin Island did not play well for Ustinov at all. He found another scapegoat—chief of the general staff, Marshal Nikolai Ogarkov—who was fired after that incident. To undermine Ustinov's position in the Politburo, Andropov brought to Moscow the party's strongman from Leningrad, Grigory Romanov. For all of us, the homemade analysts and habitual consumers of political rumors, it was an unexpected move. Romanov symbolized the most anti-intellectual kind of party apparatchik.[2] In the inner sanctum of the Politburo, Romanov was given the portfolio of secretary of the Central Committee, responsible for the military-industrial complex. That was the end of Ustinov's monopoly of the two key positions inside the Politburo and the Ministry of Defense.

---

2. The legends of Romanov's authoritarian kingship in Leningrad reached Moscow from time to time. According to these stories, Romanov cynically suppressed the local intelligentsia. The personality cult he developed in his Leningrad party estate reached unprecedented dimensions. Some of the episodes were so bad they were almost anecdotal.

For use at the wedding party of his daughter, he ordered the precious ancient pottery from the national treasure, the Hermitage Museum. People bitterly joked: of course Comrade Romanov had the right to do it. He was the successor to his namesake, the Romanovs—the last Russian Czars, whose home was the Winter Palace, currently the Hermitage.

Andropov's elevation of such an anti-intellectual as Romanov was difficult to explain. Andropov, who was considered a very smart politician, was known to flirt with artists, writers, and scientists. However, if his decision had been intended to undermine the role of Ustinov, he succeeded.[3]

In the meantime Romanov did not hurry to jump into the professional business of the military-industrial complex. My friends in the Defense Department of the Central Committee told stories that their new boss was spending most of his time conferring with regional party secretaries influential in the overall picture of the Central Committee. It was up to them to decide in the most critical period of Soviet history who would be the next general secretary of the Communist party.[4]

The first actual intervention of Romanov in the space policy domain was the drafting of a report on the civilian and commercial part of the space program for the Politburo. While most of the facts he reported regarding how poorly Soviet space assets were used for practical nonmilitary purposes[5] were absolutely true, the main idea of his initiative was again to undermine the role of Ustinov, who kept the whole space infrastructure under his unchallenged control.

Romanov's initiative in no case demonstrated that he was eager to change the rules of the game played at the top of the Soviet military-industrial complex. It was clear that he was using his current position in the Politburo as a springboard for the next promotion. But he, too, was a marked man.

Gorbachev's hour was approaching. On being appointed general secretary, the very first action he took in the internal political struggle inside the Politburo was the elimination of Romanov in the spring of 1985. While official party statements on Romanov's resignation referred to health reasons, rumors circulated about his excessive abuse of alcohol. Even if that were true, it was not against the established culture of the

3. Before that time, Ustinov was one of the three top influential leaders of the country—along with Brezhnev and chief ideologue Suslov. The first two had died and now Ustinov was no longer the omnipotent godfather of the whole military-industrial complex.

4. Everyone remembered the dramatic dismissal of Khrushchev by a vast majority of votes from the regional party bosses, and the plenary meeting unexpectedly called by Brezhnev.

5. For example, space telecommunications, meteorology, and remote sensing of the earth.

ruling elite. In the atmosphere of growing corruption, nothing brought party comrades closer to each other than the raising of vodka glasses and the making of long toasts that expressed personal faithfulness to each other.[6]

In such an environment, Baklanov was a very rare exception, I remember. The new minister of the Ministry of General Machine Building didn't much like drinking parties. At least the middle-class functionaries in the Defense Department of the Central Committee hinted at that. In the Soviet party hierarchy, it would have been equivalent to being handicapped—being unable to use a crucially important human dimension in establishing the necessary connections and ties. It was against the whole culture of party camaraderie and, to some extent, even against a kind of soul-to-soul conspiracy that was easily established, especially during the hours-long sauna parties—the new fashion of the late seventies and early eighties.

Indeed, how can a person be trusted if one refuses to be part of the narrow circle of confidants? One might easily be the wrong person, as demonstrated by a proclivity to hide secrets from comrades. I don't know how much Baklanov's sobriety actually undermined the confidence and authority he had among the ruling elite. After all, the nation was close to the introduction of prohibition in May 1985. But in the meantime, during the last agony of the stagnation period, Baklanov made a very smart tactical move. He appointed as the first deputy in the ministry a man who was a genuine party breed and one-time party secretary of the town of Kaliningrad—the stronghold of Korolev's space empire near Moscow.

6. The different degrees of friendship that emerged as an outcome of such parties essentially boiled down to one simple yet eternal question. "Do you respect me?" drinking buddies in a similar state of inebriation would ask each other. The beginner's degree of respect and friendship had to be attested to by drinking "bottoms up" in one gulp. Unfortunately, for people like me, the quantum measure of the "gulp volume" was a regular shot glass filled with vodka. However, the highest degree of friendship required real sacrifice and the readiness, if necessary, to finish the party on one's "bottom," under the table. In an ironic way, friendships that had been tested in combat and tempered in the trenches under fire were replaced with solidarity under the drinking table. However, it was done so precisely in the spirit of an old Stalinist warning, "If you are not with us, you are against us." That's how the most important alliances were established in the party politics of the late stagnation period.

Baklanov—"Big Oleg," as he was quickly nicknamed—kept sending his deputy Oleg Shishkin, "Little Oleg," to sit at the time- and health-consuming drinking parties. Perhaps this type of duty, fulfilled by Little Oleg, represented the ultimate form of friendship and loyalty.

In January 1984, I had to perform as toastmaster at Velikhov's fiftieth anniversary. A quick glance at the gathering immediately confirmed that virtually all of the military-industrial complex had attended the celebration. The long elaborate toasts reflected Velikhov's contribution to the defense programs. Half-jokingly, ministers mentioned in their speeches SDI and high-power lasers, the area that had brought Velikhov a secret Lenin Prize and the respect of the military-industrial complex for his achievements in designing them. But by that time Velikhov, decorated already with the highest national award, the title Hero of Socialist Labor, had firmly launched himself in the front line of opposition to SDI. His friends from the military and industrial sectors did not know yet how much Velikhov's efforts were to eventually undermine their interests in getting budgets and providing a stable and secure program for themselves.

During one of the breaks in the toastmaking, Little Oleg took the empty chair next to me. He was obviously in a condition appropriate to soul-to-soul revelations.

"My minister [Big Oleg] and I are planning big changes in the space program. We are going to reject the overcautiousness of Afanasiev [the Big Hammer] and throw away the blinders our space program has artificially imposed on itself."

I knew what he was talking about. Baklanov was lobbying for quicker implementation of the Soviet shuttle program *Buran*, despite the sloppy economic performance of the shuttles on the American side.[7] However, Big Oleg's ultimate goal was to find a use for the new superlauncher *Energiya*, which was intended to deliver more than a hundred tons into Earth orbit. According to their vision, it would open new horizons for the Soviet space industry. Such heavy payloads delivered into orbit would require astronomical sums of money. Someone had to justify them by arguing for such a heavy booster. At the culmination of Little Oleg's soul-searching I heard, "We wonder if you are with us. . . ."

7. It was still two years before the *Challenger* disaster.

Quickly I figured out the first part of the question: who were "we"? There was no doubt that "we" was Big Oleg, at least.[8] The second part of the question dealt with the role I could play in such company. Science per se could never represent the biggest investment in the space program. Even in our wildest dreams, space scientists could not imagine the planetary robots or space telescopes becoming the principal justification for space projects.

I was sure that the hidden part of Little Oleg's message was related to a new round of overly expensive political shows in space and, quite possibly, the growing appetite of the defense industry in military space, perhaps even in SDI. I didn't like the company inviting me or the goal they were pursuing.

The lobbying that had been launched by IKI at almost every level to promote genuine space science had already borne some fruit for the new projects the institute was developing. The next plan, to be launched the following year, was to encounter Halley's comet in deep space. I doubted that anyone could undermine this project, which was politically protected by heavy international involvement and obligations. After all, the timetable we had to follow on this project was determined by the astronomical clocks of the Solar System, not by the clocks of the Soviet party system. We knew the party system was doomed by stagnation and internal corruption, unless a miracle were to happen.

8. Most possibly the shadow of Romanov was behind Big Oleg. The rumors I heard from my friends in the Defense Department of the Central Committee spoke of the active bridge building between Romanov and Big Oleg.

30 / Gorbachev

➢

If only a miracle could rescue the Soviet system from imminent collapse, then Mikhail Gorbachev was sent by Providence. This view was widely accepted at different gatherings, from the meetings of intellectuals in academic circles to the small informal sauna parties of midlevel apparatchiks. Even the secretaries of the regional party committees, the silent majority behind the plenary meetings of the Central Committee, regarded the arrival of Gorbachev with the inevitable shake-up of everything.

I first encountered the name Gorbachev in the early seventies, while helping Artsimovich and Keldysh. Sometimes I was asked to go on special missions to the northern Caucasus, where the Academy was engaged in the construction of the two biggest Soviet astronomical instruments, an optical telescope and a huge dish for radio astronomy. The construction site was part of the largely agricultural region of Stavropol, where Gorbachev had by that time reached the level of party boss number one. I even have vague memories of seeing him a couple times when the academy bosses had to pay visits to local party feudal lords, asking for the support and supervision of our construction projects.

These encounters took place a few years before Gorbachev was launched into the all-union orbit of the Central Committee as a nonvoting member of the Politburo in 1978. He had been brought there by Brezhnev to take responsibility for the national agricultural and food program. Many people in the country were briefly attracted by the fact that such a young man had been introduced into the headquarters of the party gerontocracy.

Without undermining the impressiveness of Gorbachev's personality and his outstanding qualities, which obviously singled him out from a crowd of regular regional party secretaries, at the same time it is undeniable that he had been enormously lucky to be party boss in Stavropol in the northern Caucasus. His location gave him a preemptive strategic advantage. The area under his command was an attractive place for the

Soviet elite to spend vacation time, enjoying this glorious spa and the beauty of its mountain landscapes. The best of the privileged sanitoriums and dachas for the top echelon of Kremlin leaders were located precisely in the Stavropol region.

One of the compulsory duties of the local party boss was to make sure that the presence of Kremlin leaders during their visit to Stavropol was fruitful and pleasant. This geostrategic arrangement gave to the young, energetic, and charming Gorbachev plenty of opportunities to prove his respect and loyalty to patrons from the Politburo—perhaps at the expense of "not least heroic feats" described earlier.

Gorbachev was launched into the Politburo at an unprecedentedly young age: he was only in his mid-forties. I remember wondering if this man could introduce a breakthrough in the steadily deteriorating Soviet agricultural scene, with its chronic dependence on regular grain and food imports. Apparently a miracle did not take place, and most of us who were rather far from the corridors of power in society quickly lost interest in this recently promoted "rising star."

The first significant and intelligible signals in relation to Gorbachev came a little bit later. Occasionally, Velikhov, Georgy Arbatov, and economist Abel Aganbegyan, whom I had known well since Novosibirsk, mentioned the name of Gorbachev as someone interesting and unusual. I soon discovered that all of my friends, and a few other acquaintances, had been invited to be regular members of different parallel working groups chaired by Gorbachev. The general understanding was that this activity had been launched with the blessing of Andropov, then general secretary of the CPSU.

The message delivered by our colleagues, the participants in these activities, was quite straightforward: something is wrong with our society and now this small group of bright men are seeking the specifics and trying to figure out how to mend them. I remember one friend of mine, an economist, claimed that when he was asked directly what kind of recommendation he would make for curing society's problems, he responded with a counterquestion: "Tell me, how far are you ready to go politically?"

In the beginning, all of these stories had not yet been reflected publicly, even in the party press. Even so, everything I heard from my friends brought a feeling of relief—that something might be done—although the debates were still far from the focus of my own attention.

At that time I was completely captured by the final preparations for the launch of the *VEGA* mission to Halley's comet. I think it was my own intentional approach to stay away from big party and state politics. There was only one exception I readily agreed upon: the issue of arms control.

In the autumn of 1982, I was invited by Velikhov to take part in private, informal talks between arms controllers in both the Soviet and American academies of science. I did not have much experience in this field. However, the topics that were discussed—the ASATs (antisatellite weapons), for instance—were much easier for me to consume because, being a spaceman, I had all the necessary technical competence. Complemented with a bit of strategic thinking inherited from Artsimovich, my competence already placed me well over the curve. So my integration within such an international arms control community, beginning with space militarization issues, was a rather natural process.

The next and even more important topic on the horizon was the potential danger of a much wider and greater scale of military operations in space—the potential use of space technology for ABM, the antiballistic missile defenses. This was an old idea which I first heard about in the late sixties in Novosibirsk when Mishin, Korolev's successor, came for a visit on his company's corporate jet from Moscow. I can't forget the first bad impression created by this bossy man who emitted the strong smell of cognac. Maybe that contributed to my first rejection of the idea of SDI, even if the very name had not yet been invented.

I was quite prepared for the discussions during the early eighties on the military aspects of space issues. However, in March 1983 at the conclusion of our second meeting in Washington, D.C., with American "arms controllers" from the National Academy of Sciences, neither we nor our American friends had any idea how close the world was to an official breakthrough in debating space-based ABM, the grand vision of SDI. Only on the plane out of Washington did we hear the news of President Reagan's "Star Wars" speech.

Upon our return to Moscow we very quickly assembled a small group of people, consisting of a few physicists and political scientists, and tried to assess what had to be done. Two major concerns were clearly seen. First, technology would have to go a long way to achieve the goal. But even more, we were worried about the strategic implications. I myself, as a plasma and controlled fusion physicist, was seriously concerned whether strategic stability would be lost. Our conceptual response to SDI was drafted by the autumn of 1983. In fact, it was not much different from the views of our American counterparts in the National Academy of Sciences.

While my colleagues were becoming more and more active in nongovernmental forms of arms control, there was no sign as to whether Gorbachev at that time had any special interest in the issue. From the moment Gorbachev was chosen successor to Chernenko as general secretary of the CPSU, my friends Arbatov and Velikhov were already considered quite influential members of Gorbachev's team.[1]

The first initiatives of the new Kremlin leader were rather far removed from the issue of heavy weapons. The first recipe he tried to cure the illnesses of the system was to embark on an intensified and accelerated path of rebuilding the economy, with an emphasis on the greater role of science and high technology. He even held a highly advertised meeting[2] in the Kremlin to formulate the most important issues before the Soviet scientific and technological community. Gorbachev's appointment of Velikhov, who at that time was vice president of the academy, as principal speaker at the Kremlin created a lot of excitement—even a sensation. It was probably the first time that Gorbachev hinted at the need for renovation and rejuvenation of the system. We took it as a sign of Velikhov's soon-to-come appointment to replace the old fox Alexandrov at the academy.

Inspired by such news, I went directly to Velikhov and offered my help in preparing the draft of what I thought was his very important speech, which asked the question, How could Soviet science cure the illnesses of the system, while it too was ill and desperately in need of radical surgical treatment? With input from several like-minded, concerned scientists, I had already prepared a very critical and, I thought, courageous draft.

I was not a participant in the meeting. But even if I had been invited, I most probably would have been unable to go because of the *VEGA* mission. However, I was deeply disappointed with the last-minute

1. Velikhov had even accompanied Gorbachev on his famous trip to England, which set up a long-lasting friendship and cooperation between Margaret Thatcher and Gorbachev.

2. That meeting, in June 1985, precisely coincided with one of the most memorable space events in my life. On that very day we introduced a balloon into the atmosphere of Venus, the first component of our scenario to later encounter Halley's comet, called the *VEGA* mission.

changes. The old environment surrounding Gorbachev in the Kremlin did not support Velikhov's original intention to deliver his far-reaching speech. His actual report was rather mild and not much different from the old stereotyped speeches built according to a simple algorithm: "We have outstanding achievements. However, there are some minor shortcomings."

The road to the grand vision of perestroika was not straight and easy. The well-publicized moratorium on drinking and alcohol seemed to many of us, from the very beginning, a futile and ridiculous move. Arbatov was so outraged with the measure that, after taking part in the Victory Day Parade on May 9, 1985, as a war veteran, he wrote an emotional personal letter to Gorbachev urging him not to sign the prohibition law. Aganbegyan still thinks that the very first serious blow to the Soviet economy, which finally led to an enormous proliferation of budget deficits, came from that ill-thought-through vodka moratorium. Even at the expense of the ruined budget, it did not achieve its main goal: to reverse the old rooted drinking habits of the population.

Soon Gorbachev's interests moved in the direction of international security and military buildup, trying to break the deadlock with medium-range nuclear missiles in the European theater. I believe that outside the understandable general strategic considerations, Gorbachev wanted the chance to reduce the future military budget as a part of his program for the economic revival of the country.

At that time, I thought the assessment of SDI, and the ideas on our potential response might be of special interest to the new Kremlin leader. The circle of people who from time to time were part of our arms control discussions related to the academy's working group, which also included Evgeny Primakov and Aleksandr Yakovlev. Upon his return from the Soviet embassy in Canada, Yakovlev took over for a brief time the directorship of the Institute of International Relations and World Economy, long considered the privileged think tank of the political leadership of the country—and competitor to Arbatov's institute (of the United States and Canada) in that employ.

Arbatov's prominence as a close advisor to Gorbachev had become quite obvious. Once he said to me, "You should meet Mikhail Sergeyevich. You would be really impressed with this man. I mentioned you to him and the useful role you could play. He said he has heard your name." I have never tried to be a courtier, mixing and mingling at important VIP gatherings, so I was unable to interpret what essentially had happened.

In September of that year (1985), I received my first official invitation to take part in a meeting with the leaders of the Socialist International to discuss arms control and international security issues. While I myself played the role of an official arms control expert for the first time in my life, the encounter between the Soviet official party delegation and the European Social-Democrats (who until very recently were called "social traitors" by the Bolsheviks) was quite remarkable, and was an indication of significant change indeed. Though these discussions did not yet represent "new thinking," Yakovlev—the future architect of it all—was with us, for the last time, as our academic colleague. Soon he was promoted to the Central Committee as head of the Ideology and Propaganda Department.[3]

Later that same autumn, together with a small think tank at the Space Research Institute, I compiled a long and well-substantiated memo containing a detailed assessment of SDI and suggestions for our counter-strategy. Arbatov had suggested that I take part in such written briefings for our new leader. There was probably no more experienced man in the country, in that genre, than Arbatov. Gorbachev was the fifth general secretary to whom Arbatov had served as an advisor. However, he said, "Don't wait for any written reply, and don't be discouraged if it doesn't come."

In mid-October 1985, about a month before the Geneva summit—the first meeting between Gorbachev and Reagan—a large group of celebrities (essentially the people who were known for their important positions in the academic world or in theater, cinema, and the arts) were invited to a meeting in the Ideology and Propaganda Department of the Central Committee. The official who delivered the phone invitation to me added in a conspiratorial tone, "Please come. Don't miss this meeting. You will witness the beginning of a very important initiative."

As ridiculous as it may sound from the heights of today's achievement in glasnost, we invitees to that important meeting were absolutely stunned by the extraordinary boldness of that promised initiative.[4] Ap-

3. We all knew that a few years before, while Gorbachev was visiting Canada, Gorbachev discovered how congenial Yakovlev's thinking was on the issue of the systemic crisis in Soviet society.

4. However, we did not yet know that Yakovlev was going to become the principal ideological architect of perestroika.

parently, all of us who received an invitation to this meeting in the Central Committee were included on a special privileged list of people who were being granted the right to meet foreigners: that is, businesspeople, politicians, and journalists who regularly asked for appointments or interviews. From now on, we were told, we were entitled to say yes or no without asking the highest permission from the authorities.

It was a thrilling sensation. In a society where everything was under strict control and tight regulation, even phone numbers could not be given to foreigners. But at last, a group of well-defined people were given a special ticket for glasnost. Out of excitement, we even forgot to ask when this right to speak to foreigners would expire. But soon it became clear that this meeting was only a prelude.

I was invited again two weeks later with a much smaller part of this crowd and instructed to be ready to depart for Geneva. On the plane to Geneva, I discovered that it was an old familiar group of my friends and colleagues, including Arbatov, Velikhov, and Primakov. We were being sent to serve as the intellectual descent of the Kremlin, a new type of advance team before Gorbachev himself and the official delegation would arrive. Our only instructions were to be attentive to any call from the press and to be ready for interviews.

For a number of days in anticipation of the real summit, we held briefings on behalf of the Soviet delegation in the big press center specially opened before the summit. Such an unusual Soviet team was new and created a lot of excitement. Each of us was in great demand. A dozen interviews a day, requested by phone, inspired someone among us to invent a funny comparison: "We are indeed the political call girls of the Soviet delegation."

But at the same time, we were proud that even in this way we could contribute to probably the most important event of the time. In parallel, we had a good opportunity, while "swimming" in a rich and diverse political bouillon, to do our own brainstorming and to produce our final assessment for the general secretary. When he arrived in Geneva he had a chance, finally, to consider our suggestions.[5]

Our assessments were compared with those that came from professionals in the general staff, the military and industrial sectors, and other agencies, I presume. No one knew at that time what would be the form of our own interaction with the arriving Gorbachev and his official

5. Journalists dubbed us "the Gang of Four": Arbatov, Velikhov, Primakov, and me.

assistants. The old culture of the Soviet elite was always secretive. Up to the last minute no one was able to schedule personal time.

Arbatov and Velikhov, as senior members of the gang, were invited to meet Gorbachev at the airport; and they later parted from us to move to the more luxurious hotel booked for the delegation. At the very last minute, when Gorbachev was already in Geneva, the official messenger hurriedly came to our hotel and requested that Primakov and I immediately join the others in Gorbachev's suite for the first important conversation. I was going to meet Gorbachev for the first time.

Gorbachev was sitting in a medium-size room at a long table. With him were Eduard Shevardnadze, the newly appointed minister of foreign affairs; Ambassador (to the United States) Anatoly Dobrynyn; Marshal Sergei Akhromeyev, chief of general staff; Yakovlev; and our fellow companions, Arbatov and Velikhov. While sipping tea from big Russian-style glasses, we members of the "advance" team had a chance to tell Gorbachev about our impressions and give our assessment. Other than the different items that might be directly related to the subject of the summit—like SDI, nuclear configuration, and so on—I thought it might be interesting for Gorbachev to hear about the certain psychological overtones he might encounter at the summit.

Seizing the moment, after a brief pause I intervened: "Mikhail Sergeyevich, judging from how the American press comments on the impressions of George Schultz after meeting with you in Moscow, you should expect a lecture from President Reagan about how naive the Russians are, and you in particular, on the issue of the existence of the military-industrial complex in America. My new friends from the American press corps heard directly from George Schultz, when he was flying back from Moscow, that you, Mikhail Sergeyevich, indeed believe that there is such an animal in the United States that could be termed the military-industrial complex."

Gorbachev turned to Arbatov and—trying to provoke him—said, "Georgy, was it not you who told me about the very existence of such a thing?"

Then I added, before Arbatov was able to respond, "Apparently the American side was amused at how Gorbachev could think so when the U.S. military budget is only about six percent, which would mean that this so-called complex could never play an important role in American political life. They think you don't understand that President Eisenhower's farewell speech was simply a joke in that respect."

Gorbachev then continued to press Arbatov, teasingly, "Yes, Georgy, you always told me about the military-industrial complex, and how it controls everything."

Arbatov reacted: "Well, it's true that the nominal military budget is only five or six percent, but you have to take into account all the industries, infrastructures, and services that are related in one way or another to the military-industrial establishment. On top of it, the military-industrial complex is very well organized in the political sense, with a strong lobby on the Hill."

Although I didn't follow what happened with respect to this particular issue in Geneva and during these first encounters between the president and general secretary, a few years later in the summer of 1989, I was a guest lecturer at the Bohemian Grove in northern California. There I had a chance to talk to the then–recently retired Secretary of State George Schultz about the exciting and memorable days of the first summit.

"Do you really think that President Eisenhower was joking?" I asked.

Comfortable and relaxed, relieved now of the burdens of sophisticated arms control negotiation that had produced such an outstanding breakthrough, Schultz answered, "Of course it was not a joke. But you Russians at that time tried to oversimplify what Eisenhower said."

I heard from Shevardnadze an interesting recollection regarding the very first minutes or hours of the summit, at the time when he, too, became a private citizen after his dramatic resignation in the end of 1990 from Gorbachev's government in protest against the mounting threat of dictatorship: "We had established a kind of ritual for opening summits. President Reagan usually came with a bunch of small sheets of paper, probably prepared in advance by his aides. He would read from time to time the quotations from Lenin or Khrushchev, things like 'We will bury you.' And we in response would quote the passages on the military-industrial complex from President Eisenhower's speech."

At the time of the Geneva summit, we had tea parties after almost every working day. The evening after the very first few hours of meeting with President Reagan, Gorbachev looked unusually tired and embarrassed. He was unable to hide the feeling of disappointment at, I believe, his own failure to impress Reagan with the eloquence and power of his arguments, especially on the issue of SDI. I think the ego he had developed during the many years of his party career convinced him that he could persuade anyone in the Soviet Union about anything. There were only a few people who did not fall under the spell of Gorbachev's personal charm and the magnetism of his verbal talent.

I myself have had the huge fortune during my life to meet and fall under the spell and influence of such outstanding and bright personalities and interlocutors as Lev Landau, Lev Artsimovich, and Andrei Budker—great thinkers and debaters who had unusually quick reactions. However, in his own weight category—politics—Gorbachev, I admit, was almost overwhelming. Perhaps over the years, as he discovered in himself an ability to convert people—to change their opinion—he developed the feeling that he was irresistible, and capable of controlling a conversation with anyone.

While admiring his talent, that of a genuine born missionary, I thought at the same time how easy it is to be mistaken in overevaluating one's own power to persuade the audience when, as a rule, it consists of your clients and subordinates.

However, at least one good thing came out of this interactive process and Gorbachev's own evolution as a leader: that is, his own ability to influence people's minds by simply talking to them. Even if he did so in a most impassioned and eloquent way, it was a sign of great progress in the political culture of my country. His approach was in sharp variance to the tradition that bosses usually adopted, who never tried to change people's genuine opinions or beliefs, but simply issued an instruction and demanded that it be followed.

Another impressive feature of Gorbachev, which eventually made him a world champion in disarmament, was his ability to learn and, subsequently, to make up his own mind about complicated issues of strategic stability and arms control. I recall an interesting remark he made to a group of his advisors in early 1986: "I bet there are as many definitions of strategic parity as we have people sitting in this room. I am ready to defend my own. Real strategic stability does not necessarily require that both sides follow each other, nostril to nostril."

I am pretty sure Gorbachev was doing his own share of homework and brainstorming on such an important, vital issue in order to avoid following blindly what he was told by the generals.

One particular episode I remember with some feeling of vindication. At a small meeting in Gorbachev's Kremlin office, an official from the

Soviet space industry, Aleksandr Dunaev, the president of Glavcosmos Agency,[6] noticed Gorbachev's genuine involvement in strategic thinking. Seizing the moment, Dunaev said, "Dear Mikhail Sergeyevich, I completely understand your concerns. Trust me. We are losing time while doing nothing to build our own counterpart to the American SDI program." I almost died from supressing my laughter. A lot of people, not only the Gang of Four, but Marshal Akhromeyev also apparently tried to influence Gorbachev not to copy the SDI program. At the time, Akhromeyev was probably the best expert on that strategic issue.

Perhaps Dunaev was simply fulfilling instructions from Big Oleg, his minister. This brief episode gives at least a clue as to why we Russians were, as many in the West thought, overescalating the anti-SDI rhetoric. In my own frame of reference, I always had in mind the potential danger that could arise if Big Oleg, Dunaev, and other similarly influential members of the military and industrial sectors in our own country were to involve us in a nonstop escalation of an SDI budget at that time, at the expense of the deteriorating strategic stability and our economy.

Speaking to a seminar of political scientists and strategic analysts in Paris, I even confessed that "if Americans oversold SDI, we Russians overbought it." Jim Hoagland, then Paris correspondent for the *Washington Post*, asked whether he could quote me. Wishing to stay on the safe side, I asked that my name not be mentioned in the final article. Reference was made to an anonymous "Russian official."

Yes, we paid too much attention to SDI, but I believe we did so for a good purpose. At least it saved the country a few billion rubles.

Looking back on that brief period of "honeymoon" with perestroika, I cannot avoid a feeling of nostalgia—nostalgia for our early romanticism and naive hopes to change the system without firmly rejecting it. And poignancy for the period of Mikhail Gorbachev, whom we thought to be the savior of the country, the reformer of the system. No one at the time (and least of all Gorbachev himself) could predict that Gorbachev had a much more important predestination, a completely different fate. It would be this savior who would inadvertently accelerate the collapse of the system.

6. Created to promote the commercial services of the Soviet space program.

31 / The Call of the Name

➢

I often wonder why psychologists have not fully analyzed what seems so obvious to me: the correlation between the fate of a person and the name he has been given or inherited. Happy parents probably never imagine the consequences inherent in making their newborn child a namesake of an heroic figure of their epoch. Think how much they have already predestined the future life of a Staliny or a Marat. Such kids automatically become hostages to their names. I believe such names lead to a continuous subconscious hammering, a constant reminder of the significance of the name they have been given or inherited. It must create a deep inner urge to emulate the feats of their heroic namesake.

With my own given name, I tried to deny any connection with my eventual appointment to lead the Space Research Institute in the Soviet Union. After all, I thought, the distance between traveling to the South Pole in sleds carried by dogs, as Norwegian Roald Amundsen did in the early twentieth century, and flights to the distant planets, driven by rocket engines of the space era, is tremendous. However, once at a meeting with spacecraft designers, I was struck by a peculiar notion. There were three Roalds in one small room; in the heart of a Slavic country, three bearers of a rather rare Norwegian name. It couldn't be completely accidental. We all must have been drawn by a subconscious desire to emulate Roald Amundsen as our role model. We were brought up to become space explorers by the simple fact that all the poles on our own planet Earth had already been explored.

With one of these Roalds—Kremnev, a leading engineer in Lavochkin Design Bureau,[1] I had already dealt on the launching of the robotic probes to Venus. But those probes, as well as virtually all the projects of the Space Research Institute that I had had to supervise during almost ten years in the chair of director, were adopted before my time. I simply had to implement them as an administrator—no doubt while subconsciously

1. The principal contractor of unmanned planetary spacecraft.

dreaming of a South Pole of my own and waiting for the call of the name.

In September 1979, the French Space Agency invited a Soviet delegation to Corsica for the regular annual meeting on space cooperation. Our hosts were strongly pushing us to launch our huge air balloon to Venus in 1983 to celebrate the bicentennial of the Mongolfier brothers' flight.[2]

Toward the end of this meeting, despite the beauty of the Mediterranean, we were exhausted. After long discussions that covered every aspect of space cooperation—from space medicine to planetary launches—the balloon to Venus became the major focus of our discussions.

At the concluding party, my friend Jacques Blamont, the chief scientist of the French space program, told me how painful it had been for the European space scientists when NASA canceled earlier plans for a joint Euro-American mission to encounter Halley's comet. We tried to assess the prospects of learning more about Halley's comet during our upcoming encounter in 1986. The comet, passing too far from Earth, would provide only a narrow opportunity for astronomers, even with modern telescopes, to learn substantially more than the previous generation of astronomers had learned in 1910.

Space astronomy, of course, could observe the comet from an orbit in the ultraviolet spectrum that is absorbed by the atmosphere of our planet. So, discussion moved to the possibility of launching such a telescope, with a Soviet booster, into orbit around Earth.

Then I said, "Look, Jacques, what about our own Venus balloon project? Why not use the ultraviolet telescope scheduled to study the cloud layer on Venus—while relay spacecraft are orbiting around that planet?" The closest approach to Halley's comet this time was Venus, not Earth. That conversation with the use of quick charts—drawn on paper napkins—was the beginning of what later became a completely new project.

Initially we could dream of nothing more than accommodating additional scientific instrumentation for exploring Halley's comet and, at the same time, keeping—untouched—the original package of the huge Venus balloon. Not trespassing on the interests of other groups of scientists, we were simply talking about making cometary science a "guest

2. Somehow it was rewarding and reassuring for us that we Russians were not the only nation launching rockets to celebrate important anniversaries.

passenger" on the balloon mission. Even that modest scenario would provide the chance to have a look at Halley's comet from a distance of 40 million kilometers—much closer than one could observe the comet from Earth. Even so, it was a rather remote view of the comet.

However, our appetite started to grow and discussions that were triggered in Corsica continued. Something new was raised almost every day, until the moment we asked ourselves, "Are we bound to staying in orbit around Venus? Could we possibly move closer to the comet?"

In January 1980, after an intensive brainstorming at IKI, we developed a fairly clear alternative option that was designed as an intentional technical and political compromise—between the original balloon mission to study the superrotation of the Venusian atmosphere and a very close encounter with Halley's comet. According to this new scenario, the initially suggested huge balloon had to be replaced with a much smaller one, capable of carrying only a few hundred grams of science, instead of twenty-five kilos of diversified scientific instrumentation. The substantial savings of energy and mass could be used to send the spacecraft, after delivering such a small balloon to Venus, in the direction of Halley's comet. The very passage near Venus would also serve a double purpose beyond simply delivering the balloon: we could use the gravitational field of that planet as a springboard to get a special kick in the right direction to intercept Halley's comet.

On the technical side, all the cards came up very well. Both missions, the original one and the cometary one, were quite compatible. The tiny gondola of the new miniballoon was still capable of carrying a small walkie-talkie–type radio transmitter that could send signals to Earth. The sensitivity of existing radio telescopes on Earth was sufficient to pick up even very weak signals from such a transmitter, the power of which was equal to the light of a simple candle. At the distance between Earth and Venus, over 100 million kilometers, it would be enough to follow the motion of the balloon as a very accurate atmospheric tracer on Venus. At any given minute in time we knew we would receive the position of the balloon with an accuracy on an order of a kilometer.

On the political side, we discovered that the double-purpose mission, while seemingly more complicated, was actually capable of saving a substantial amount of money. For instance, it no longer required four Proton launchers. With only two launchers, we were able to perform all the space maneuvers.

However, even though the new scenario had such very clear advantages, we had to reverse the momentum that had been built in favor of the original project over the course of several years. We had to do it in both countries, in the Soviet Union and in France. I remember that at the appropriate moment I took color slides and a slide projector to give a talk to a large crowd of Smirnov's assistants at the headquarters of the Commission on Military-Industrial Issues. I knew I should first focus my lobbying on the lower-level staff. After all, it would be up to them to promote the project were they to buy my argument.

Speaking eloquently and passionately, I thought, I was close to winning their support. Then one of the participants of that party stood up and said, "I am ready to agree. It sounds extremely impressive, but have you prepared a soft bed to protect yourself from a rough landing?"

I had never felt such a drive to succeed. I knew it was a moment of truth for the institute, for our space science, and for me, who had already spent quite a few years in the chair of director. Forcefully, I answered, "You are producing multiple warheads, ICBMs. What you see here is essentially a three-warhead rocket: two deliveries to Venus—a small balloon and a lander to the surface; and the third to encounter Halley's comet. What we are going to do is simply use the military technology that you developed."

I still believe the use of such an analogy was a decisive argument in converting this audience into my ally. Somewhat later, at the meeting of the presidium of the military-industrial commission, I went even further by trying to tell Smirnov and the ministers of the military and industrial sectors: "To deter the Americans, we are building hundreds and thousands of silos for ICBMs. There they could be perceived only as dummies. Don't you think that such a spectacular launch, shooting at several targets in space at the same time, separated by tens of millions of kilometers, would not only be impressive but also serve as intelligent deterrence?"

The big bosses kept their silence. Later, after the meeting was over, one of them dryly said to me, "Do you hope to deter Pershings this way?"

Even with all the right cards, I still had a major problem: who would be the "icebreaker" for the project? Keldysh had already died. Ustinov, sitting in his military headquarters, had lost interest in philanthropy for the space program. Suddenly, I decided for myself that this was precisely the moment when we needed to change the paradigm. Instead of making important decisions behind closed doors, I would try to promote it openly, changing the minds of every audience in the Council of Ministers and in the Central Committee of the CPSU.

One of the first stops in my campaign was a seminar held in the headquarters of the main scientific and engineering establishment. The State Committee on Science and Technology assembled more than a hundred atmospheric scientists and chief designers. For the first time during this seminar, I believe Boris Petrov, as the head of Intercosmos, the Soviet umbrella organization for space cooperation, understood how strong the support from the scientific community was and decided to lend his authority to the new project.

In the meantime, Blamont had to overcome his share of complications in Paris. The engineering group of the French Space Agency and their supervisors were far from expressing enthusiasm. Of course, they had to change their initial technical scenario from the big to the small balloon. And real perestroika was needed to do it. Time was pressing. We had to make all the preparations and integration of the final payload in only four years—much less time than the average planetary mission would require.

Most of the French space experts and officials openly expressed their skepticism. Even the support and sympathy we had for the project from the then president of the French Space Agency, Hubert Curien,[3] could do little to reverse the situation. Our opponents openly criticized us for "adventurism" and warned us of the risk associated with potential failure of the project. My answer was, "Don't worry. Together with Blamont we are ready to accept the final punishment in the spirit of international cooperation. While Blamont would be sent to a Siberian gulag, I am ready, in exchange, to accept imprisonment within the walls of the Bastille."

One of the biggest blows to the project came with the final refusal of France to support the balloon part of the project, or to take any responsibility for any aspect of it. For a brief moment, I felt defeated. Under these circumstances, the only chance for the project's survival was to reinvent balloon technology in Russia, almost 200 years after the Mon-

3. Curien soon became even more influential as minister of science in the French government.

golfier brothers. Fortunately, the original spark of the idea ignited a shock wave of enthusiasm, which was propagated throughout the space community in the Soviet Union. That was how we found a team of contractors among Babakin's former pupils.

At the same time, IKI launched an appeal to the international space science community to come up with proposals for scientific instruments to study the comet at close range. Even at such short notice, with already less than four years before the launch, we got proposals to participate from nine countries.[4]

In the beginning, there was not much hope of having participation from the American side. The most we could count on was strong support from the Jet Propulsion Laboratory (JPL) in Pasadena, to employ NASA's deep space network to follow the signals from our tiny balloon on Venus and then from the Halley's comet probe. Later these plans became an essential part of the whole *VEGA* project. They enabled us to navigate not only *VEGA* spacecraft but, with much greater precision, the European cometary probe *GIOTTO*, which was to meet Halley's comet at a distance of 500 kilometers. This particular part of the overall project, bringing together Soviet *VEGA* spacecraft and *GIOTTO*, soon was given a special boost under the Pathfinder project.

While we had no hope of bringing American scientific instruments on board the *VEGA* spacecraft, even when it was an absolutely obvious scientific opportunity for Americans to encounter Halley's comet,[5] we were realists and understood that the then-current state of Soviet-American relations was less than favorable toward us. The rhetoric of the Cold War had reached its peak in the early eighties.

Hence, I was rather surprised when one day I received a fax from my *VEGA* colleagues in Germany. They called my attention to a new proposal coming from the University of Chicago's space scientist John Simpson, who had proposed to study dust particles from Halley's comet. Apparently the original request was to put such an instrument on the *GIOTTO* spacecraft. However, as tiny and light as the new American

4. I remember at one of the first international meetings of our team in Budapest, someone suggested we invite the legendary Rubick, inventor of a popular toy cube. His ingenuity, they said, could be of great help for the project. With a smile, our Hungarian friends presented participants Rubick's latest toy—The Tower of Babel. "Our" multilingual team was already strong enough to reject superstitions. We spoke the common language of international science.

5. NASA had no plans to encounter the comet.

instrument was, *GIOTTO* was already overloaded. The question was, could *VEGA* accommodate such an experiment? I asked if Professor Simpson would be ready to meet my representatives. The answer came soon.

"Yes, he is waiting for an invitation." This time I had to be very serious and attentive—this had obviously gone beyond polite correspondence. On the Soviet side, too, we were far from ready to have such a discussion. I could not even imagine how much time and effort it would take to get the formal approval of our own authorities, especially in the extremely hostile atmosphere after the downing of the Korean airliner over Sakhalin Island. But time was pressing, with less than two years remaining before the launch.

Rather than try to navigate the tempestuous seas of the official bureaucracy, I decided to take the responsibility on myself. Thanks to the courtesy of my Hungarian friends, they agreed to invite John Simpson for a confidential meeting.

Now I believe that if I had not taken such unilateral risks, the *VEGA* project would never have materialized. The Soviet regime was stagnating and virtually paralyzed by the escalation of bureaucratic inefficiency. Even if I had decided to follow all the formal instructions, nothing by this time was working anymore. My open and hidden supporters in different Soviet governmental offices told me quietly, "Don't ask us. We don't even know how to get approval for it. Just do it. If you succeed, nobody will accuse you. But if you fail . . ."

I knew there was no way to fail. We were doomed to win. Our excitement and enthusiasm conveyed itself to almost every institution, including the midlevel bureaucracies. I think a lot of people personally felt they had been denied the right of self-expression. They felt tired of the hypocrisy of society and the futile bureaucratic fuss. I think they generally wanted to be helpful. I think even the KGB security officers, supervising the activities of IKI and related aerospace companies during this time, were excited by the project.

Simpson's race to join the comet team was not without hurdles, too. He told me that a seemingly menacing interagency commission interrogated him in Washington. "Are you sure Russia won't steal our technological assets?"

"To avoid it, we at the University of Chicago intentionally built our instrument only from components purchased at the Radio Shack store on the next corner. Besides, the Russians promised never to touch our instrument. It will be in a black box for them."

But an inquisitive voice interjected, "What if they would secretly break into it at night?"

"It would set them back ten years," replied Simpson, winning the ticket to the comet.

Jointly with the European Space Agency (ESA), NASA, and ISIS, the leading space science institute in Japan, we formed a special panel to coordinate the activities of the space mission to encounter Halley's comet. It was called the Interagency Consultative Group. The name was vague. But for us the substance was more important, and the vagueness of the wording saved us from the bureaucratic bonds in our own home countries.

For the first time during my long administrative career as director, I felt no split between the necessity for management and an active scientific life. Both sides were indispensable for running such an exciting and complicated project.

The leading experts in Soviet orbital dynamics—people who precalculate the trajectories of spacecraft—were skeptical about the possibilities of predicting the orbit of Halley's comet with enough precision. I invited an independent and younger group of experts in celestial mechanics to do the peer review and create the competition. It worked.

The Soviet space industry was unable to mobilize itself quickly to build a special platform to point the telescope at the nucleus of the fast-moving comet. So IKI appointed young Czechoslovakian engineers, and they did the job nobody in Russia believed could be done in such a short space of time.

With the Hungarian team, we found the formula for providing our spacecraft with elements of artificial intelligence capable of finding the tiny cometary nucleus, the surface of which was obscured by dust jets that were formed by evaporated ice. In virtually all these technical and engineering problems, I was most intimately involved. If such an undertaking happens only once in life, it was my chance. And I knew I had to take it.

The barons of the Soviet space industry at this most difficult and sensitive moment, however, were not on our side. Every time a new technical question arose, they tried to simplify the scenario of the mission or avoid accepting the responsibility.

IKI had to intervene and be in charge of almost every new and nonstandard component of the project. The big bosses, the general designers, perceived us in IKI as kamikazes. Indeed, it *was* a kamikaze-type mission in many ways. The *VEGA* spacecraft on its approach to the comet had to encounter streams of dust particles that would hit it with a velocity exceeding eighty kilometers per second. To complete the mission and send the data to Earth, the spacecraft had to survive only for a very brief period of about an hour. However, even the tiniest dust particle, with a mass not exceeding a milligram would in principle be capable of disabling the spacecraft by cutting its ability to communicate with Earth.

The question was how to protect the spacecraft against such a deadly weapon of nature, comparable maybe only with the nonexistent exotic technology of the beam weapons fabricated for *Star Wars*. Upon completing the first quick-look analysis, based on my own background in plasma physics,[6] I figured that the instantaneous pressure at the brief moment of impact could jump to 100 megabars (millions of atmospheres). I suggested that we possibly involve the best experts in such an apocalyptic science, who were working in Chelyabinsk-70, the place I had luckily avoided. I approached my old classmates who were still serving the nuclear devil in that weapons laboratory, to see if they could help make the final assessment for the dust threat problem and suggest how to build the shield. Evgeny Avrorin responded immediately: "Yes, we are going to help you." That small exercise was perhaps one of very few diversions from the deadly business that occupied their lives in the bomb industry.

The *VEGA* project was moving quickly. One particular phenomenon became a by-product, not anticipated at the very beginning of the project. The very buildup of strong cooperation between enthusiasts created tremendous pressure on the whole bureaucratic establishment and its stagnating spirit. We were pushing for greater flexibility, quick response, openness, and the ability to send more and more people abroad when necessary. We were not simply pushing the limits; we were demanding that they be stretched. It was amazing to feel that such drive worked. The very launch of the *VEGA* spacecraft in December 1984 was the first disclosure on Soviet television of the Proton launchers.

6. After all, at such a speed the impact of the dust particles on the spacecraft envelope would immediately produce microexplosion that would emit a cloud of that very plasma substance.

The guardians of the old dying system tried to resist and defend their trenches. Spectacular Proton launchers took two *VEGA* spacecraft to vast interplanetary space out of the huge flame and smoke that accompanied the ignition of the first stage. Then authorities insisted that no details on stage separation in the boost phase should be given to the foreign audience or the wide Soviet public. We survived that ridiculous insistence with a joke we told our friends and participants from the nine nations: "Remember, for the very first time in your life you have assisted the launch of an absolutely incredible booster that does not separate in stages. It simply works as a unique monoblock."

The balloons to Venus were delivered in June 1985. We missed the bicentennial of the Mongolfier brothers' flight by two years, as well as the bicentennial of the great French Revolution. So what!

The moment the balloons started the fifty-hour-long journey as tiny man-made objects, following the atmospheric streams on Venus, General Secretary Gorbachev was launching, maybe unconsciously, another revolution in our own country: perestroika. On that day in June, he talked about the need for acceleration in industry and technology. I do not think Gorbachev realized at this time that, without much fuss and support from the establishment, the scientists and engineers of his own country supported by the international space community were *in fact* revolutionizing the approach to international cooperation and openness.

I have never felt such intense anticipation as when we waited for the first signal of the *VEGA* balloon from the distant planet Venus. I was sitting in the control room of IKI, next to French ambassador Ramon, when the intercontinental phone call came from California informing us that the big crowd of scientists and engineers at the JPL control room were getting a signal from NASA's distant dish in Australia. The deep baritone of General Lew Allen, then director of JPL, congratulated IKI and all participants with the first success. Knowing how risky space exploration is in general, and especially such sophisticated projects as *VEGA*, I had the feeling that the Good Lord was on our side.

The final stage of our spacecraft—after Venus—was moving toward our assumed encounter point with Halley's comet. The comet, still very far away, was seen mainly through the optics of large telescopes. But we knew it was going to meet us on March 6 of the next year, 1986, almost nine months later.

Ted Koppel's "Nightline" was first to realize that the space orbits of *VEGA* and Halley's comet would intersect each other at the very hour of the show. I was asked whether IKI would be willing to take part in the live broadcast. There was not much support in Soviet official circles for such unprecedented publicity, which violated the long-established practice of hiding everything until it is successfully concluded. But early in the morning of March 6, 1986, I sat in the control room at IKI with Carl Sagan and other friends and colleagues. We were in an elevated emotional mood. The image of Halley's comet, and the scientific data sent to us every few seconds by the *VEGA* probe and quickly processed by computers, were displayed on the large screen of the monitor. Koppel asked us questions about the comet. Sagan and I commented on the first data. It was an unprecedented dialogue, an interview that preempted the advent of glasnost.

During precisely these same hours, Gorbachev was addressing the Twenty-seventh Party Congress. I suspect he had no idea about the *VEGA* project at all. At least not until the president of the academy, Alexandrov, upon receiving my phone call, approached him and told what had happened just a few minutes ago in deep space. However, the even deeper irony was in the fact that Gorbachev undoubtedly had no remote idea or hint that this particular apparition of Halley's comet, intercepted by Soviet spacecraft, delivered the celestial omen that communism was doomed.

Three days later, the second spacecraft, *VEGA II*, approached the comet somewhat closer and at a different angle. That was the moment when the call of my name came true: *VEGA II* sent back images of the South Pole of the comet's nucleus.

32 / The Revenge of the Comet

➢

When *VEGA I*, *VEGA II*, and then the *GIOTTO* spacecraft flew by the comet, it was already on its ascending trajectory from the sun. Soon it would fade in the distant outskirts of the Solar System for another seventy-six years of oblivion. Unlike the previous apparition in 1910, fewer people were to see it with the naked eye. In practical terms, to observe the comet one would have to fly to Australia or any other point in the Southern Hemisphere. Even so, it probably would have been difficult to see the comet by simple means. Despite this, I thought it was absolutely essential that I pay tribute to the comet and try to see it from Earth, not just through the *VEGA* spacecraft's imaging cameras.

A friend of mine, the head of the astronomical observatory in Kiev, told me that there would be a time at the end of April (1986) when the comet could still be seen in the sky. But this could be done only through telescopes. He invited me to come to his observatory to bid farewell to the comet.

It was a good chance to go to Kiev, and we agreed that I would come with my family to stay during the weekend before May Day. We were looking forward to enjoying the early Ukrainian spring and the beauty of Kiev. This season is always pleasant, especially on the high hills surrounding Kiev where this small observatory is located. I looked forward to the trip and to seeing my colleagues at the observatory.

Suddenly a day before we had to leave, my friend called me from Kiev and said, "Maybe you should reconsider your trip and not come to us now."

I said, "Why? What's happening?"

He said, "I cannot speak openly, but there has been an event at a nuclear power plant here."

Being a nuclear physicist myself, I knew that such plants had been run for several decades, and of course there could be some small accident but, I imagined, nothing more. Then he said, "Okay, I'll risk it, I will tell you. There was a *major* accident at the Chernobyl nuclear power station.

You should ask your colleagues in Moscow if they would advise you to come or not."

I said, "Please stop kidding me. Nothing could be that dangerous eighty kilometers from such a station." There had been nothing in the press, but that night I switched on my shortwave radio to hear what was being said abroad. The whole world was already full of talk about the Chernobyl accident. Apparently Sweden had been the first to detect the radioactive fallout. But still there was nothing in our press.

I decided that I should call my friends at the Kurchatov Institute, and they confirmed that it was better to be very cautious and prudent. Their last word to me was: "Okay, if you are all set to go and you promised to be there to give a talk . . ."

In fact I had promised to give a talk to astronomers on the *VEGA* encounter. "If you have promised . . . at least go alone. Do not take your family," they advised.

But my wife and daughter were very stubborn on the subject. They, too, had been anticipating the nice trip. Having no such scientific experience, they said, "If you can go, why shouldn't we go with you?" They resisted any idea that it might be dangerous. We thought, after all, Kiev was a city with 2 million inhabitants. Even if something *were* wrong with the nuclear power station, the residents were sentenced to stay there forever. Why shouldn't visitors stay a couple of days? So, we packed our bags and headed for Kiev.

When we got to Kiev, tension was already evident in the city. After leaving our things in the Ukrainian Central Committee's hotel, we were taken to the observatory. No official data had been issued, although there was a statement in the press about a certain accident at Chernobyl. My friend kept informing me about the radiation measurements taken onsite by the observatory. This was the third day after the accident, and the level of radioactivity in Kiev was surprisingly low compared to what had been reported by the Swedes. That was because the direction of the wind and the atmospheric circulation pattern were very favorable. Nothing had gone in a southerly direction yet.

Forgetting this accident for a while, we took pleasure in looking at Halley's comet. It seemed like a very strange, weak, and diffused object—very different from the really spectacular image on our computer displays, as viewed by the *VEGA* or *GIOTTO* cameras at their closest approach. After my talk, I was invited to the May Day celebration. Later the soccer team of the observatory, a group of young men, invited the older guys to play a friendly match. I made the sad observation that even among the veterans I was the oldest.

Although the game was a good one, the old players understandably lost. Soccer is a tough game that requires power and strength and constant training. But somehow I had a chance to deliver a penalty stroke on behalf of my team when the opponent, the youngsters, violated the rules in the vicinity of their goalkeeper. I don't know, maybe the honor of delivering the penalty stroke was given to me as the oldest member of the team. It was such a joy to play soccer. The only sadness was that we were swallowing the dust on the soccer field that was certainly, by now, at least partially contaminated with Chernobyl radioactivity.

Our last night in Kiev, we could already feel the wind changing, and on our last day the level of radioactivity in Kiev started to rise. It was such a sad farewell party with my friends. We joked that the toasts we were raising to our friendship, and for our cooperation in future projects, would also play an important role in keeping our bodies resistant to the radiation. But of course Kiev was in a state of panic. In a few hours, iodine, a very important preventative medicine against radioactive iodine, disappeared from all the pharmacy shelves.

I think all the official authorities were taken by surprise, and very simple recipes for the people to avoid overexposure and stay inside came too late. Only later were all the instructions given.

After I returned to Moscow, I got a telephone call from a friend of mine at the Kurchatov Institute. He asked me if we had gotten checkups after coming from Kiev. I said, "Don't be ridiculous. What kind of checkups are there for people who spent only two or three nights in Kiev?"

He said, "You better be ready, I'm sending my team with radiation monitors, just to gauge your situation."

Before long, a technician with a monitor came. We were rather surprised to find out that the shoes we had worn in Kiev were giving off enormously high counts of radiation. So we put them in a plastic basket and let the technician take them back to his institute.

Yevgeny Velikhov, busy with his Kurchatov experts, was really seriously alarmed that the Chernobyl situation was still completely out of control, possibly with the graphite inside what used to be the reactor burning at a high temperature. Heroic efforts were made to stop this process. Attention was focused on assuring that liquid uranium would not slowly make its way through the layers of graphite and ultimately through the thick layers of concrete plate supporting the body of the reactor. It was imperative that it not reach the huge water reservoir in the basement. The contact of high temperature material with water could have been fatal in the final, worst-case scenario.

Many people shuttled between Moscow and Chernobyl: experts on radiation, chemistry, and material sciences, as well as the former contractors and the military. Our institute volunteered to be part of the computer simulation efforts to consider different potential scenarios of what might happen to the reactor, especially if the melted blob of uranium moved down. Portions of the computer-animated movies produced in our institute were taken to the Vienna meeting of the International Atomic Energy Agency as important materials on the potential dangerous scenarios surrounding the Chernobyl tragedy.

I think that at this most intense period, when everyone felt a sense of emergency, many enterprises and institutions in our country tried to contribute to easing the difficulties. I suggested trying one of the techniques we had developed at the Space Research Institute—to remotely operate the probes that had been used in the hostile atmosphere of Venus. The payloads could be delivered into the open crater of the reactor, where they could survive for a while sending messages by radio.

Our team went there for a few days. However, unfortunately, our device was lost. While hurrying to fly over the disaster site, the helicopter pilots, whose responsibility it was to drop the capsules, missed the open mouth of the reactor. Despite our concern, no one blamed the pilots. They were the kamikazes of the Chernobyl rescue operation, navigating their helicopters through jets of radioactive smoke.

The Chernobyl efforts also utilized another technology from our space program. One of the companies, an offspring of the tank industry, sent a descendant of the lunar rover Lunokhod. This planetary rover, modified for future robotic missions to Mars, was used on the roof of the reactor building. It did some important work in cleaning the roof of radioactive waste.

The connoisseurs of history claim that each stellar apparition of Halley's comet every seventy-six years has brought a warning of forthcoming great events. I have never considered myself superstitious. After all, to be superstitious goes against a physicist's own rational thinking. In this respect the physicists identify themselves with Niels Bohr. During an exemplary episode, an interviewer came to speak with Bohr and discovered a horseshoe on the door of the great scientist. Expressing his astonishment, he said, "Professor, are you superstitious?" With an ironic smile Bohr responded, "Not at all, but I believe the horseshoe brings luck even to those who are not superstitious."

Despite my commitment to the rational thinking of a physicist, I could not stop feeling a strange shiver at the thought of the comet's sacramental omen throughout the history of humankind. Its fiery tail in the sky was considered a precursor to the great battles that forced powerful empires to break up and disappear from the map. The comet inflicted devastating epidemics, terrible earthquakes, and floods that destroyed flourishing civilizations.

We were joking bravely at our gathering on the eve of the *VEGA* encounter with the comet. "Even if it were true in the past, this time everything is going to be different. At last human genius is able to design and launch a spacecraft to intercept the mysterious comet." However, I could hardly sleep the last night before the first encounter on March 6, 1986.

While the *VEGA I* spacecraft quickly approached the giant fireball surrounding the comet's nucleus, a few miles from the Space Research Institute inside the Kremlin Palace of Congresses, Mikhail Gorbachev prepared to deliver a speech to conclude the Twenty-seventh Party Congress of the CPSU, the first congress he convened as general secretary. I caught myself wondering if he had ever thought about the omen of the comet—what it might mean for him and the fate of his political party.

The excitement of the encounter and the overflow of scientific data completely absorbed my time and thoughts during the next few days. I didn't have much time to think of politics or omens until I came back to earth and received a message that Gorbachev had invited me, as part of a small group of scientists and aerospace officials, to brief him on the encounter with the comet.

At the beginning of the meeting he asked the first question in a teasing way: "How did you manage to encounter Halley's comet on the very last day of the Party Congress?"

"Mikhail Sergeyevich," I responded in the same style, "this was pure coincidence. We have the perfect alibi; such timing was not intentional. After all, the orbits of comets are decided upon in heaven."

Gorbachev immediately took this opportunity: "That means that God is with us."

Who could predict, at that time, the true message of the comet!

The Twenty-seventh Party Congress, whose timing coincided with the cometary encounter with *VEGA*, did not bring relaxation of the old practices of one-party domination. While all the doors of the Space Research Institute were opened for the international community, the party congress was unable to institute glasnost and openness in the country. In fact, an important member of the Politburo, the chief of the KGB, Viktor Chebrikov, complained about the infiltration of an alien ideology and way of life with the help of photocopiers and VCRs.

The intelligentsia took the results of the congress as a great disappointment. We thought that maybe Gorbachev had been unable to overcome the resistance of the old guard. Even the election of a new ruling body in the country, the Central Committee, was practically a victory for the apparatchiks when people like Yakovlev were still kept in the second echelon. (He had not yet been promoted to the Politburo.) The only hopeful exception, however, was the promotion of a newcomer named Boris Yeltsin. He was made a nonvoting member of the Politburo and the head of the Moscow city party.

Yeltsin did not keep us waiting too long. The very first steps he took in Moscow were quite bold, not only rhetorically, but through direct action. He rejected much of the old style of party control and micromanagement, as well as a substantial fraction of the old cadre.

However, the majority of newcomers that had been brought to power by Gorbachev were his old cronies—his party comrades with a long track record of proof, as he probably thought, of personal loyalty. One of them, Ivan Murakhovsky, who used to be the second party man in Gorbachev's estate in Stavropol, was given the job of revolutionizing and rebuilding Soviet agriculture. If it was Gorbachev's own narrow professional area of competence to become a model for perestroika in society, from the very beginning very few people had any illusion that scenario would work.

Gorbachev and Murakhovsky wanted to strengthen the collective ownership of the land and the system of collective farms. The missing element they intended to bring was the cross-breeding of that system with an industrial approach. Their attempt to rescue Soviet agriculture was based on a new magic word, *AgroProm*, which is the abbreviation of the agricultural-industrial complex. Perhaps deep in his heart Gorbachev thought that there would be no chance of the country's feeding itself unless this complex were able to establish itself in the life of the country—at least in the same way the military-industrial complex had done. The approach, however, was doomed from the very beginning.

In addition to the Chernobyl disaster, in the autumn of 1986 there was a terrible catastrophe on the Black Sea. The huge passenger cruiser *Admiral Nakhimov* collapsed within a few minutes, after being hit by a cargo ship in view of the well-illuminated Black Sea port and in perfect weather and visibility.

The whole country mourned the victims of the sea tragedy, an occurrence that led to a real sense of pessimism and foreboding.

Black humorists invented a story about a captured CIA agent. His interrogators tried hard to attribute the country's disasters to the conspiracy of a foreign power intent on bringing down the system.

"So, Chernobyl was a CIA plot. Admit it."

"Oh no, no."

"What about the Black Sea incident—would you deny the role of your agency?"

"We had nothing to do with that, you must believe me."

The poor captive CIA agent started to cry. The interrogator brightened the lights further and leaned forward menacingly.

"Okay, okay. If you have to know, the CIA had nothing to do with Chernobyl or the accident on the Black Sea. But we invented Agro-Prom—that was our idea."

I believe that the popular anecdote, as painful as it was at the time, played a crucial role in the eventual resignation of Comrade Murakhovsky, and the decline of the Gorbachevian concept of AgroProm.

The man who most symbolized resistance to the sweeping democratic changes in society was Yegor Ligachev. All of us were afraid of this alter ego of Gorbachev and caretaker of the party *nomenklatura* and all its appointments. Since early perestroika, we knew that in the inner kitchen of Gorbachev's cabinet, Ligachev was the main opponent of all the ideological innovations suggested by Yakovlev and the bold pragmatic steps proposed by Yeltsin to rebuild the party network. In that complicated environment, the shock of Chernobyl became an important test of Gorbachev's personal qualities as a national leader in the revolutionary transition period.

For more than two weeks after the Chernobyl accident, Gorbachev kept silent. This silence effectively helped official propaganda downplay the graveness of the nuclear disaster. But for most of us it was simply unexplainable. After more than a year of continuous nonstop sermons on the imminent change in our society, we were shocked by his handling of the tragedy. Perhaps Gorbachev was paralyzed as Stalin had been during the first weeks after the outbreak of World War II. Such comparisons came to the minds of many of us at that time. Nevertheless, we tried to indulge Gorbachev, blaming his performance on misinformation and cheating, which would have reached him through official channels. Such powerful special interests close to the Kremlin quickly identified the reactor controller, at the lowest end of the nuclear industrial hierarchy, as the culprit—because he had pushed the wrong button. That's how the *nomenklatura* and the nuclear establishment tried, not without a temporary success, to protect themselves.

In the Space Research Institute, while helping the authorities simulate the events leading to the disaster with the help of black market Western computers, we knew the attempt to make the controller the scapegoat was a blatant lie. However, a dangerous cloud started to gather around the father of the failed system.[1] But we knew that the hours of the old fox Alexandrov, the official supervisor of nuclear energy engineering and science, were numbered.

My own relationship with Anatoly Alexandrov was, even before this disaster, rather strained. I had my own reasons to believe that as the president of the Academy of Sciences, Alexandrov didn't like the Halley's comet group from the very beginning. Our *VEGA* project had minimal support from the president of the academy, and the briefing with Gorbachev a few days after the encounter was arranged largely by Velikhov.

1. The father of the failed nuclear reactor design, not yet of the country's failed system.

Nevertheless, the *VEGA* project by itself fell into a category that assured it would receive a special governmental decree, with a list of decorations for the participants. By midsummer 1986, a draft decree prepared by the academy and the Commission on Military-Industrial Issues was submitted for final confirmation by the government. IKI waited for publication of the news in the press.

But the comet played one more joke. Some officials in the Kremlin discovered that the awards could not be reported in the open press since the *VEGA* project had been supervised by the military-industrial commission. Several dozen people from IKI were to get different types of decorations, as well as those from other institutions. I was given the title Hero of Socialist Labor.

However, this new complication with the confidentiality of the decoration made me laugh. The mission had been discussed and assessed on television! I asked my friends to please keep it secret that I had been decorated for the mission to Halley's comet, because the whole project should be considered classified spying on the comet.

Arbatov did not think the situation was so amusing, and he took the initiative to negotiate with the government. He explained the ridiculousness of the case to Ligachev, who at the time was substituting for General Secretary Gorbachev. The deadlock was eliminated and the decree was soon published in the newspapers, and I was able to accept "legal" congratulations.

Rumor came from the supervisors of our institute that the presentation of decorations would be made by Yeltsin at a special meeting. It seemed appropriate enough to us that this dramatic new star who had appeared on the horizon of Moscow would be the man to present our decorations. As the omen of the comet might have foretold, it was Yeltsin who would seal the fate of the Communist party and the Soviet Union.

33 / No Perestroika in the Academy

➢

I was asked to "run" for the Supreme Soviet at the by-elections held after the sinking of the *Admiral Nakhimov*. In a wave of indignation and protest, the deputy for the Odessa region had been fired and I was recruited to replace him. This invitation came perhaps in recognition of our successful venture to encounter Halley's comet, as well as my political contributions to the process of arms control and international security. But society was turning more and more attention toward the international political aspects of perestroika, to the problems of reshaping society. This interest was accompanied with a lot of discussion on the recent political history of our country. I thought that it would be illogical during such an active political rethinking, which was becoming a major theme of the country's life, to continue only as an expert on technical aspects of arms control. With more and more interesting revelations and disclosures in our national life, I witnessed a growing interest in the people, especially among the intelligentsia, to engage in politics. I gradually became involved myself in such rethinking.

One particular area that required reconsideration was the state of the Academy of Sciences, as well as the life of the scientific community within the framework of the academy. It was natural that I should speak out on these topics, where my professional life had been conducted for more than thirty years. Although it was perhaps inevitable that eventually I would be drawn into these issues, there were a few additional reasons why I became radicalized and then active in this debate.

First of all, the tradition that came from Artsimovich was there. He had been deeply involved in the field of international security. Second, I felt guilty about how things were developing in the Academy of Sciences. There was a general belief in some camps of the intelligentsia that the Academy of Sciences, and science in general, was evolving as an island of conservatism, in contrast to other groups of intellectuals who were working toward perestroika much more rapidly. Among the most conservative groups, the Writers' Union and the Academy of Sciences were still kept under the control of the old guard.

From time to time there were critical articles in the press, written by younger scientists who depicted the academy as a conservative elitist group bent on keeping its privileges during a time of renovation. This doubt and pessimism about the role of the academy only enhanced my sense of shame because, in a certain sense, I had been part of a process in the autumn of 1986 that had led to the election of the new president of the academy, backed by a very important member of the Politburo, the man who was thought to be the second man in the party, Yegor Ligachev.

In October 1986, I received a telephone call from Comrade Ligachev: "Would you be able spare some time to talk to me?"

It was an invitation to the old familiar building of the Central Committee. I had no idea what the subject of the conversation would be, but it was clear that it must be something very important. I tried to figure out the topic of discussion, at least generally, in order to be psychologically prepared. When I arrived a few minutes before my appointed time, I noticed in one of the waiting rooms the vice president of the academy, Kotelnikov. Obviously, the staff did everything to put us in different rooms. I figured out that it had been done intentionally. It was clear to me that some kind of important campaign was being waged here. Most probably it had to do with changes in the academy.

When I entered Comrade Ligachev's office, he greeted me very warmly and said, "How is your life going in Moscow? Do you need any help?"

I realized that I was being treated as a Siberian comrade, who shared with Ligachev his beloved part of the country. He even hinted at a closeness between us.

"Thank you very much," I replied, "everything is okay, I don't need any help."

He then moved to the issue: "You know that eventually the academy will need a new president. There are many concerns expressed by different people about the state of the academy, and many suggestions have been put forward in letters from a number of academicians. The Politburo has recommended Guri Marchuk as the future president of the academy. So, what is your reaction to that?"

I was amused with such an approach. They were asking me my reaction *after* the Politburo had already made their recommendation. I remarked, "Of course, Guri is a capable person, very quick. The only problem might be that he's too quick, that he tends to make decisions without thorough consideration. He might be too light-minded."

Comrade Ligachev still smiled in a friendly manner. "You know, this is why you and I should help him."

This part of the discussion was over, and somehow my neutral stance was interpreted as an agreement to speak for Marchuk. While I was uncomfortable and unhappy about such a prospect, the subject had been cleverly changed before I had an opportunity to psyche myself up for a refusal.

Ligachev was interested in my assessment of how the space program was developing, and he assured me that, if necessary, he would be ready to intervene in any way. At the end of the conversation, he expressed again the hope that I would be able to say a few words about Marchuk a few days later at the academy's formal meeting, when the election procedure would take place. He wanted me to speak specifically at the party group meeting, which usually preceded any official academy business. That was the way decisions from the top were implemented.

The events were arranged as a "blitz" campaign while Velikhov, then the candidate of the liberals, was not yet in Moscow. He had accompanied General Secretary Gorbachev to Reykjavik. Probably he was to be the last one to discover that an election was being held.

The meeting of the academy started with the ritual of the party group, in fact assembling more than 80 percent of the people from the list of academicians. The meeting was chaired by Vice President Kotelnikov, who probably had had a similar briefing with Ligachev. He said that Academician Alexandrov was asking for retirement.

"He spent eleven years in this post, and we have to be thankful for his contribution to the development of the academy. Now let me give the floor to Comrade Ligachev, who agreed to be with us on this very important day."

I did not expect that my opinions would be considered so important. But I was asked to take the floor as the next speaker. I said a few words: that I used to work with Marchuk in Novosibirsk and that he was always full of energy and was always quick. Then I thanked him for the support he had given the *VEGA* project while he was chairman of the State Committee on Science and Technology.

Now I had reached the trickiest part of my talk:

"Of course, some could ask how we should react to his controversial performance in that committee. I would suggest the interpretation be that he kept his powder dry for the future." That was the end of my speech.

Formally I had offered some supportive words about Marchuk, but the formula I used probably did not please him. Its true meaning was understood by many. Some privately congratulated me, immediately after this part of the meeting was over. "Oh! You were clever; you said nothing good about him, but at the same time you did not violate the rules of the game."

I think that was part of the tension that developed later in my relationship with Marchuk, and in the attitude of the new president toward IKI. However, I have to confess that my sense of guilt did not disappear. I severely chastised myself for not being brave enough to reject the pressure from Ligachev and the Politburo to become an official speaker at the meeting.

Moreover, I should have openly opposed such a recommendation for president of the academy. But all of us were products of the epoch; and that epoch had not yet begun to decisively change. Even well-intentioned liberals like myself were not yet psychologically ready for open disagreement with members of the Politburo.[1]

However, my conscience continued to bother me, and my first critical remarks on the academy's internal affairs at that period were made in the spring of 1987. I had the chance to take the floor a couple times to criticize the academy's attitude and assessment of its president on the state of computer science and computerization. I was particularly unhappy with the euphoric mood in the official statements of Marchuk on how rapidly and successfully we were moving toward the development of supercomputers. Marchuk had been among those responsible for painting an inaccurate picture of our computer state of the art.

Many of us were also critical of the Marchuk presidency for the strengthening of the echelons of the bureaucracy. It was reminiscent of the stagnation period. The newly appointed president had brought with him scores of a fresh younger breed of apparatchiks, who created a kind

1. Even Yeltsin's courageous and, at the same time, desperate precipitation—an outcry against what he thought was a sabotage of perestroika—took place a year later, in October 1987.

of buffer group between the academy leadership and the scientific community.

I was not alone in such criticism. Quite a few other people criticized this development. But all of the dissent simmered outside the presidium or at its plenary meetings. The idea to express these views in the open press did not yet occur to me.

I was approached by *Moscow News* to comment on the last academy elections. This was how my very first public statement on the current shape of our scientific community and academy life was published in the press. Over the course of the next several years I would write and be interviewed for countless more statements.

The appointment of Marchuk as president of the academy was a painful disappointment, not only for Velikhov. Many in the academy took it as selling out to the Ligachev-Marchuk mafia.

If the intellectual community was considered the potential stronghold of Gorbachev's social and political reforms, the appointment of Marchuk was a betrayal of the very notion of perestroika. An old prophecy of Andrei Budker—"Wait until Marchuk becomes president of the academy; it will be a real disaster"—soon materialized.

Both Velikhov and the new president of the academy were competing to bring the computer age to the nation. Unlike space, in the field of computers and microelectronics, the Soviets trailed behind the world standard by about fifteen or so years. Even in the nuclear area, the Chernobyl disaster could have been attributed to a lack of appropriate computerization in designing and controlling the nuclear reactors.

The most striking indication of Soviet backwardness was the absence of domestically made supercomputers. The supercomputer was considered a strategic attribute, the lack of which was inexcusable for a superpower. We knew how the government and its military-industrial commission issued their top-level confidential decrees and instructions to accelerate the work on national versions of the supercomputer. The old established technique of smuggling wouldn't work in this case. It was too sophisticated a technology and too well guarded for the usual black market practices.

As a complete humiliation to the Soviet leadership, the news broke about the development of the first Chinese supercomputer. Velikhov, the responsible official in the computerization business, launched an investigation on the reasons we had not yet attained this goal. I was appointed chairman of that committee. After a few months of work and briefings with leading computer designers in the country, I was ready to share the findings and see Marchuk.

When I met him in his office, I wasn't given the opportunity to even open my mouth before the president of the academy declared, "I myself did all the necessary research and came to a final assessment. I don't need to keep your commission working anymore. We have arrived at our conclusion."

I could only guess at what kind of conclusion he himself had reached. But a couple months later we all found out what he apparently had in mind. In April 1987, Gorbachev accepted an invitation to speak to the All-Union Congress Komsomol leaders. In the course of this speech, Gorbachev surprised everyone by declaring that at last the nation, engaged in successfully implementing perestroika, had been able to build its own supercomputer.

Velikhov, who attended the speech, was almost paralyzed. He remained speechless. However, a shock wave of indignation propagated through the professional circles of the computer industry. The story became such a scandal in intellectual circles that I even heard about a group of young mathematicians at Moscow State University who were planning to write a collective letter to Gorbachev. It was unbearable for me that there could be such blatant cheating, incompatible with glasnost, "new thinking," and everything else we were trying to implement.

I assumed that Gorbachev had not known what he was saying. There was no institutional mechanism that could reject such fabricated reports—except for the Academy of Sciences. But as I was to discover, the academy had played an integral part in the cheating in the first place.

I decided to precipitate events and write a letter to the general secretary: "Dear Mikhail Sergeyeivich, young computer experts in the country are bitterly complaining that even Gorbachev can be cheated. . . ." Then I added the story of the ill-fated commission on the supercomputers I had chaired, and I enclosed a draft of that commission report. Arbatov made sure that my envelope was delivered to Gorbachev himself.

For a few weeks all was quiet from the government on this subject. However, widespread rumors told us that the government proceeded in the established confidential manner to decorate "the heros of the national supercomputer group." I was surprised to learn that the highest award for this "achievement" was the Lenin Prize, given to the chief designer, Boris Babayan. It was ironic, I thought, that Babayan, as a member of my recent commission, had beaten himself on the chest trying to explain why he had failed in designing a supercomputer. In the second group of award recipients were the closest associates of Marchuk, including his son.

I can psychologically understand Gorbachev's tremendous urge to accept the fabricated reports on the supercomputers. He desperately needed at least one indication of the success of perestroika. But he can't be absolved of responsibility for it, particularly as we had confirmation that he had received my letter and a copy of our commission report. Arbatov asked Gorbachev about my letter and his opinion, and he said, "Sagdeev may have an opinion, but there are scientists who have a different opinion."

Who these other scientists were was not a secret at all.

Yuli Khariton, my old friend, still running the nuclear installation Arzamas-16, also complained about the supercomputer cheating. After all, who would know better? Arzamas-16 would be one of the most important priority customers of supercomputing, with their ever-growing need for what is called numerical simulation.

While visiting Gorbachev a few months later, Khariton told me he conveyed his own testimony: "Mikhail Sergeyeivich, concerning the supercomputer, Arzamas has its own experience with the Elbrus system. We know very well that it is no supercomputer at all."[2]

This story Khariton told made me furious since I was aware that there had never been a hint of supercomputing here. Nevertheless, no one was punished, no one recalled the medals. There were no reports in the press either. It was as if everyone silently forgot the supercomputing cheating. It was with great sadness that we understood, finally, that Gorbachev was more comfortable with the assessment of Communist party apparatchiks than the evaluations of real scientists.

It took a good deal of time before the truth finally came to Gorbachev. In 1989 Yakovlev told me how Gorbachev had reacted when the subject of computers was raised at a Politburo meeting at the end of 1988. After hearing another "glorious report" claiming an even more powerful

2. Elbrus was precisely the system named as the highest national achievement that had brought Babayan, its principal designer, the Lenin Prize.

computer, Gorbachev apparently said, "Wait, don't hurry with a claim. First verify if it is true. Computers are not tractors."

For me there was also an ending to the story. In 1988, I participated at one of the academy's gatherings, at which I called on the academy to identify those responsible for the cheating. Marchuk frowned, smelling genuine danger. After the meeting, he asked me to come to his office for a confidential conversation.

It was then that he essentially confessed that he had been forced to participate in the charade by Lev Zaikov, supervisor of the military-industrial complex—Gorbachev's appointee in the Politburo.

34 / "Trust But Verify"

➢

The propensity to deliver—nonstop—enthusiastic reports on Soviet "brilliant achievements" reached even Gorbachev's most favored area: arms control. A crowd of attention seekers entered the verification area with almost the same fervor that they displayed in promoting their new armaments. After verification became an important issue, ideas of how to "verify" proliferated at a breathtaking rate—thanks to Reagan, who gave new life to the Russian proverb "Trust but verify."

In 1987, while discussing a potential 50 percent reduction in strategic nuclear arsenals—the early version of what is now called the START treaty—negotiators encountered a serious obstacle of what to do with sea-launched cruise missiles (SLCMs). For decades SLCMs were considered short-range or even tactical weapons for sea battle. Technological breakthroughs provided them with a strategic delivery system that enabled them to fly at a low speed, compared to ballistic missiles. At very low altitudes they are difficult to detect with radar, especially when they are equipped with stealth technology. The potential to fly such cruise missiles at large distances—up to a few thousand kilometers—was regarded as a serious potential threat.. Unlike ICBMs, SLCMs were much more difficult to verify because of the resemblance to their nonnuclear siblings. Because of that, until late 1987 there was a deadlock in the Geneva negotiations.

One potential way to detect the very existence of nuclear weapons on ships would be to use radiation detectors. Any kind of nuclear warhead, even when it sits quietly in position, emits a certain amount of radiation. The products of residual nuclear reaction, neutrons and gamma rays, penetrate the envelope of the warhead and get into the atmosphere. Since the 1930s, physicists have known the main features of radiation transport. The atmosphere shields such radiation at a distance of about fifty meters.

One could bring such techniques of verification to the negotiating table. But detecting radiation at a distance of a few tens of meters would not impress arms controllers. It would be equivalent to on-site inspection on land. At sea, it would require a very close approach to the ship under suspicion. In this case it would be too close for comfort.

A few months before the Washington summit, at the end of 1987, the director of the Vernadsky Institute of Geochemistry, Valery Barsukov, came to Velikhov to announce that his institute had developed a way to sense gamma rays from a distance of ten kilometers. The promise of this miracle was so exciting that Velikhov, without checking with his colleagues in his own Kurchatov Institute, promoted this idea at the highest levels. Obsessed with this "revolutionary breakthrough" in arms control, and the prospect that it would provide the necessary assurances for the completion of the START treaty, Velikhov reported it to Gorbachev. A few days later Velikhov mentioned it to me in a casual way, as if it were a done deal.

I jumped in immediately: "Yevgeny, how can it be possible? We physicists know everything about gamma quanta and how they propagate in the atmosphere. It is a derivative of the very basic facts of physics."[1]

I was shocked by the spell that Velikhov was under. If Velikhov, a professional physicist and an academician, was so excited, one can imagine what would happen to the arms controllers, diplomats, and big bosses. It was impossible to explain.

For the big shots in administrative jobs, it is always a challenge to be able to distinguish between real science and a bluff. When Velikhov's colleagues at the Kurchatov Institute heard the news, they were terrified that somehow it was already too late, that the genie was out of the bottle. However, I believe that at that very moment, Velikhov would have still been able to stop the propagation of his idea among the highest circles. Alas, he was not brave enough. He probably thought that the idea would be forgotten. But unfortunately, he did not take into account the principal law of paper movement: "Once written, always alive."

A couple of months later, just after the summit was over on December 10, Gorbachev gave a final press conference at the Soviet embassy compound. Velikhov and I did not attend the press conference but were instead headed for New York. When we got to the train at Union Sta-

1. Such radiation would be dissipated in the atmosphere over a distance of a few tenths of a meter.

tion, the press conference was already over and our Princeton University colleague, Frank von Hippel, told us he had just heard the stunning announcement on the radio: Gorbachev had disclosed that there was a technical way to solve the SLCMs verification problem.

Velikhov was obviously shocked. He could not believe that his idea had been introduced in this way. The statement at the press conference made no direct mention of distance, but the language sounded as if an important breakthrough by Soviet scientists had removed all problems with verification.

American reaction was quick and predictable. American experts thought that the announcement was ill-advised, but they still wondered how to interpret the statement. I also heard a joke attributed to some American scientists: "Maybe the Russians have discovered a new fundamental law of nature that is yet unknown to Americans." It was said with great sarcasm.

What damage this did to the international reputation of our country, in the midst of these important negotiations! These exaggerations and overstatements could eventually harm the credibility of our position at the negotiating table.

When I got back to Moscow, the first thing I did was to call in Oleg Prilutsky, a space physicist and arms control expert in my institute. Together we wrote an article for immediate publication. It was a very simple and straightforward analysis, based on known facts about the radiation from fissile materials. The final feasible distance we came up with indicated that radiation could be detected only as close as 100 meters—even with the most sophisticated detectors. I believe our essay played a sobering role in removing the exotic interpretations that originated after the press conference. But the impact of this article on internal Soviet affairs was even stronger.

Barsukov had been responsible for a very poorly thought-out concept, made particularly bad by the fact that he is not an expert at all in this field. This topic required some knowledge of nuclear physics. Moreover, he had a deputy at his Vernadsky Institute, a physicist who regarded himself as a nuclear expert. This fellow had apparently used the naïveté and overriding ambitions of his boss.

Unfortunately, within a very short period of time, momentum for this idea had grown considerably inside our arms control circles. Officials responsible for the Geneva negotiations continued to hint about a kind of miracle in the hands of the Soviet nuclear scientists. The professional scientific community, cautious about promising such miracles, was criticized for its orthodoxy and conservatism.

We joked that the magic SLCMs verification was based on a kind of nuclear telepathy. But we did not laugh when the Soviet navy moved a dozen of its nuclear surface ships and submarines to verify Barsukov's ill-thought-out fantasies. I phoned Yuli Khariton and Evgeny Avrorin, our top nuclear weapons experts, and suggested a joint approach to stop this craziness.

The three of us sent a letter to the Central Committee. Even Alexandrov, the old fox, cosigned it. However, we never got an official answer. After all, Barsukov, before taking the chair of the institute director, spent ten years as head of the sector inside the science department of the Central Committee. And apparently not in vain.

Our letter from four academicians was taken to the office of Big Oleg for consideration.[2] His reaction was probably predetermined by comradely ties of class solidarity with Barsukov. I was not much surprised when Baklanov, accompanied by an escort of staff members from the Central Committee, made a significant visit to the Vernadsky Institute. In order to make sure that this expression of political support was properly noted, a brief communiqué appeared in *Pravda* the next day. Though the verification technology was not mentioned directly, the message of support came through clearly.

A few days later I met "Little Oleg" Shishkin who had been in the entourage during the visit. He tried to avoid raising the touchy subject. According to him, the goal had been to familiarize Big Oleg only with planetary science: "If you are interested, we can organize a similar visit to your institute." I did not try to pursue such a visit.

As we say, "Science has its own table of ranks." In October 1988, we had elections for the ruling body of the academy, its presidium. There were a few vacant chairs. One was given to earth sciences, and Barsukov was competing against some of his colleagues for the position. When his candidacy came under discussion, official leaders in geology cited his important contributions. Then an older man stood up. It was Alex-

2. Baklanov had finally succeeded in stepping into Uncle Mitya's job as Central Committee secretary of the entire military-industrial complex.

androv, the former president of the academy. He spoke as a former director of the Kurchatov Institute, one of the professional experts in radiation physics. He strongly rejected the candidacy of Barsukov because some of his work in "unidentified applied areas" had been completely wrong. After Alexandrov left the podium, Khariton added his voice to Alexandrov's assessment.

One of the academicians, a distinguished geologist, interjected, "You are speaking in such vague terms that we cannot understand what kind of errors Barsukov apparently committed."

Khariton replied, "I cannot speak in front of this large audience because it is a classified issue."

That created an embarrassment: a very serious accusation was delivered publicly without a proper explanation. At that moment I raised my hand and asked for the floor. "As a man who is not burdened with state secrets," I said, "I can spell out openly what happened." And then I told the whole story.

At the ballot, Barsukov was roundly defeated. However, to complete the story, a few months after the smoke had settled—somehow by itself—[3] the vice president of the academy, Yuri Osipyan, visited the Vernadsky Institute. Osipyan told me that Barsukov had given him a brief outline of the current important research topics. There was no more about SLCMs verification, but one of the labs apparently was involved in new, even more exciting research on remote sensing of the AIDS virus.

While I was fighting the battles against this type of cheating, the verification fervor penetrated even my own institute. One of the planetary scientists at IKI, Gleb Istomin,[4] came up with a proposal to study the impact of chemical warfare on vegetation. He was promised a big budget by the Ministry of Defense if he would design a technique to verify the use of nerve-paralyzing substances by the potential enemy. Despite the promise of ample funding, there was a consensus among members of the institute to decline such a contract. Outside of moral considerations, we had a practical counterargument: it was not an area of the institute's competence.

3. That is, without official statements.
4. Recognized for his earlier studies of Venusian atmospheric chemistry.

Unfortunately, however, our inventor did not know how to take no for an answer. He was indefatigable. Within a short space of time, he had persuaded a number of deputy ministers, generals, and even some academicians to write me letters demanding that the institute become their contractor for this research. These approaches relentlessly continued until the chemical warfare siege reached its culmination in the autumn of 1987. It came with fire from the biggest gun: Defense Minister Marshal Dmitri Yazov sent an angry letter to Marchuk expressing his indignation that Sagdeev felt he could simply ignore the requests of important officials on issues related to national security. I am sure the president of the academy was delighted to dispatch this letter to me, requiring an explanation.

In order not to waste time, I produced a draft letter to be signed by Marchuk himself. In it, Marchuk was to advise Marshal Yazov to look for experts in the sensing of chemical weapons from among those who make them. They, after all, would be the most competent people to do this research. I did not see the final response, but rumor had it that Marchuk had changed my draft. I don't know what he said to Marshal Yazov, but there were no more complaints and no more letters.

Then one evening in the spring of 1988, I had what became my very last conversation with Valentin Glushko. He phoned me to convey an urgent request for me to sign a document that would approve a space project, which he declined to describe on the "unprotected" phone circuit.[5]

Glushko apologized for such short notice. Without giving me a chance to learn about the proposal, he nevertheless still expected me to sign the document. "Don't worry," he said, "the project already has the support of academicians Alexandrov and Khariton. You know you can trust them. Go ahead and sign." And then he added, "Or you can ask for an explanation from your collaborator, Doctor Istomin."

The next day the institute had to spend a great deal of time on this issue. First, we had to find out what this project was all about. My findings overwhelmed and surprised me. The Soviet military thinkers, paranoid about the threat of SDI, had created a worst-case scenario. What if the whole rhetoric on the American side was nothing but a giant smoke screen and the real intention of SDI was to deploy hydrogen

5. This type of confidential conversation was precisely intended for the *kremlevka* line I lost a few years ago.

bombs in space, masqueraded as innocent SDI assets? Of course, it would be the gravest violation of an old treaty banning the deployment of nuclear weapons in space. Nevertheless, they argued we should have a technique to detect such a violation. Istomin's proposal (using the same chemical sensors) was supported by Glushko for space tests on the orbital station *Mir.*

The biggest surprise came when I did a quick-look analysis that revealed that the authors had made a computational mistake, exaggerating the sensitivity of the sensors by a million times. The following, more detailed assessment made by my working group concluded that indeed the proposal's error was off by a hundred thousand times. It was more than enough to reject the project as a fraud.

Glushko did not come back to me about my refusal to endorse this project. Not long after our phone call he had a stroke and died soon afterward.

I was curious to know how my friend Yuli Khariton had appeared as a supporter of this misguided project. The explanation was very simple. Khariton explained to me that he had been told that the project had emanated from IKI. "I agreed to sign it because your institute is so highly respected," he said.

Alas, not everyone else had such respect for us. About a year later, a Russophile newspaper—*Literary Russia*, famous for its national chauvinism—blasted me as an evil genius, bent on undermining a genuine Russian inventor (Istomin) and thus damaging national security. I was not the lone target of the attack. Alec Galeev, who succeeded me as director of the Space Research Institute in 1988, was declared an enemy, too.

In the summer of 1988, the Academy of Sciences had to undergo the reelection campaign of the directors in many of its institutes. According to the then–newly introduced charter for such administrative posts, age limits had to be observed: sixty-five years for mortals (nonmembers of the academy), seventy years for simple immortals (regular academicians), and seventy-five years for the elite among the immortals (members of its presidium). Clearly, I failed to meet any of the criteria that would release me from my directorship. I was ready for a change, and anxious to leave my institute on the high note it had enjoyed with the *VEGA* mission. But I knew if I were to resign at this point, I would have to raise this issue before the reelection meeting.

I went to my immediate supervisor in the academy, academician Prokhorov, chairman of the Department of General Physics and Astronomy. We had a long conversation in his office, I had reminded him of how long I had been in the chair of director (over fifteen years). I told him how difficult and painful it was for me to deal with all the bureaucracy, and with the aerospace industry and the military-industrial complex. In essence, I told him how much I wanted to go back to science. I left Prokhorov with a warm feeling—he completely shared my views. The next step was to open the electoral campaign in the institute.

I called a staff meeting and explained my decision. The staff of the institute were not particularly surprised. I had written extensively on the problems facing the academy, and I had also called on the system to adopt rotation as a central plank of institutional perestroika. At our staff meeting I said, "Now I will leave you. You have to start discussing the potential candidates."

In about two hours, I was invited back to hear the verdict. They said to me, "We have spent two hours discussing all the potential candidates, but before we proceed we would like to ask you a question: would you agree to stay for at least two more years, so we could prepare for a smooth transition to the new director?"

My response was a bitter, short speech explaining my views of democracy. "After all," I said, "democracy and the freedom to choose should be extended also to the directors; they should also have freedom of choice."

Then there were practical debates about who might succeed me, and several names were mentioned. Three were finally drafted as candidates. All three were bright scientists working in different fields of space science. Only Galeev was a member of the party, but at this particular time its importance was greatly reduced. I believe the question of party membership was not taken into any consideration. As a matter of fact, the party bureau had no more influence on these elections and did not express its opinion. It was a first.

The campaign within the institute put me in a delicate position. Galeev was running, and as a pupil of mine since the age of nineteen, I had to be careful not to be biased. I made an official statement that I would neither participate in the electoral campaign nor give any recommendation to the institute. I said that all three were outstanding scientists; they all deserved to be director.

Rashid Sunyaev, one of the candidates, supported Galeev from the very beginning and withdrew his name from the list of candidates. Finally, Galeev became the director. I don't think he was particularly happy with such a promotion. He knew very well how much more his life would be burdened with all of the responsibilities and duties of such a post.

When he came to me for a practical discussion after he was elected, he remarked, "Do you remember, Roald, what you told me once a few years ago? You were in a philosophical mood and said, 'I wonder if I would ever recognize the moment when I would come to believe myself to be truly indispensable as director'?"

This was typical psychological transformation. If you were to talk to the younger bosses, younger directors, they would always speak about the crucial importance of rotation. They would be happy to be replaced by someone at a certain time. But if you were to talk to older men in these positions, they would rarely admit that rotation is a good idea.

In my case, at least, I wanted to leave my post of director while I still thought it was a good idea to do so. The institute I had tried to build for more than fifteen years was on solid footing. We had a budget for a few good projects, reasonably tolerable relations with our supervisors in the government, including VPK, and a decent "armed neutrality" with the space barons from the military-industrial complex. In an odd way, it was they who wanted the space spectaculars to impress Gorbachev and boost their empires, which were suffering from lack of support for the Soviet version of SDI. No doubt this is why Big Oleg turned his appetite toward Mars.

Mars was still on the minds of the scientific community, but the idea to implement the manned mission to Mars was picked up by Big Oleg and sold to Gorbachev as an immediate goal to serve as a form of conversion of the military-industrial complex.

Gorbachev made it a topic of considerable visibility at both the Washington summit in December 1987 and the Moscow summit in June 1988. Before the state dinner in the Kremlin, during the Moscow summit in 1988, Gorbachev took Ronald Reagan through the courtyard of the Russian czars. While passing by the memorable relics of ancient Russian "super" projects that were never used, Gorbachev passionately spoke about another grandiose project capable of matching the achievements of our great Russian ancestors. He told Reagan that the Soviet space industry already had the technology to fly a joint mission to Mars. No doubt he was still emotionally involved in the anti-SDI rhetoric and thought that Mars could divert the American military-industrial complex from SDI.

Later that evening there was a state dinner honoring Reagan, and the advisors of both sides were invited to attend. When it was my turn to be introduced to the guest of honor, Gorbachev seized my arm and said, "Mister President, this is the man who is promoting the flight to Mars." I had the funny feeling that Gorbachev's words struck some chord of curiosity in Reagan. As if to underscore his apparently successful start to his Mars public relations campaign with the American president, Gorbachev added: "Academician Sagdeev has friends and colleagues in America who share the same vision of a joint flight."

Then Gorbachev turned to me, as if looking for help with a few names. But before I could react, he went on: "Carl Sagan."

In a fraction of a second I could tell that something had clicked the wrong way. The guest of honor appeared to lose interest in the subject immediately. Gorbachev apparently didn't understand that there was not a great deal of political compatibility between Ronald Reagan and Carl Sagan. Gorbachev would have had better success using the name of General James A. Abrahamson.[6]

It was not simply the SDI challenge that drove Gorbachev's initiative on the grand space venture. With the economy in disarray and backwardness in virtually all areas of modern technology, he did not have many trump cards in his hands. The past several decades' nuclear achievements had been discredited by Chernobyl, and some skeptics had even concluded that the Soviet Union's entrance into the nuclear age had indeed been premature.

Space and rocketry still seemed to remain the stronghold of the country. So why not make it the role model for the rest of the national industry, instead of using it for political window dressing as it was during the Khrushchev and Brezhnev regimes?

Speaking in Prague a few months before the summit, Gorbachev complained bitterly, "It is a paradox that we can successfully launch *VEGA* spacecraft to encounter Halley's comet, but we are unable to produce decent washing machines or vacuum cleaners."

6. At that time the head of the SDI office in the Pentagon.

"Big Oleg" Baklanov, the minister of the space industry, felt that his moment was coming. He prepared a huge show for Gorbachev at the Baikonur launch site, and a group of general designers—the nobility of the military-industrial complex—paraded before Gorbachev with their posters and mock-ups, all of which reflected their irrepressible appetite for a bigger space budget.[7] Gorbachev was impressed with the scope of the program and the grandiosity of the Soviet shuttle, *Buran*, which was ready for launching. No one told him that it was obsolete and doomed to fly only once after about 20 billion rubles had been spent on it. Nor did they tell him that the superbooster *Energiya*, while capable of delivering more than 100 tons into orbit, had no purpose—except SDI or a mission to Mars. Nevertheless, the space industrialists performed well, and the Gorbachev meeting launched Big Oleg into the highest possible orbit: Gorbachev appointed him secretary of the Central Committee for the military-industrial complex. Baklanov's dream to succeed "Uncle Mitya" Ustinov and Grigory Romanov had been realized, and the road to political lobbying was now wide open. If Big Oleg wanted a big Mars project, it was not to please the scientists. Meanwhile, we were struggling for support for a much more modest Martian mission called Phobos.

Despite my resignation, I hoped to stay in touch with space science. Liberated from everyday responsibilities of directing an institute, I kept only one major task there—to run the Phobos project, which I thought would be my swan song.

7. IKI had not even been informed about this event, which took place in May 1988.

35 / The Failure of the Phobos Mission

➢

The idea to go to Phobos evolved from our early debates on an unmanned exploration of Mars. Searching for ways to diversify our approach, we turned to the neighboring moons of Mars, Phobos and Deimos—the two closest moons in the vicinity of Earth excluding our own. The interest in choosing Phobos as a target for the next mission was based on its analogy with asteroids—and the idea of astronomers that Phobos might be a captured asteroid. It was the very dark appearance of Phobos, so unusual for the planets, that hinted at asteroidal connections. Asteroids belong to a class of small bodies in the Solar System, and most asteroids are probably comparable to the nuclei of comets. Being very small, they can keep the early, primordial signatures of their material without undergoing the complicated planetary cataclysm of melting.

By the end of the seventies, we had decided to bring the proposal for such a mission to Phobos and Deimos to the attention of the general public. Support was almost unanimous for such a mission, which would replace the sample return from Mars as the next step.

The original plan was to land spacecraft on Phobos. However, engineers in the space industry were extremely worried about some unpredictable and unexpected environmental problems on such a small celestial body. Instead, they suggested that the spacecraft approach Phobos at a very short distance—twenty to thirty meters. From there it could pick up a sample of pristine material by using technology similar to that used by whale hunters. The small sample of Phobos material would be "harpooned" and delivered to the spacecraft. All analyses would be done inside the spacecraft, based on quite a few sophisticated proposals developed at the Space Research Institute.

After a while, this idea also was abandoned. An engineering group came to the conclusion that this operation would be too risky. The only option they could provide was to fly a spacecraft slowly over the surface of Phobos at a distance of up to some tens of meters. What a pity to be so close to the surface without being able to land and pick up a soil sample. However, it was almost like science fiction to discover there were exotic technologies that would enable us to sense and retrieve data on the intricate chemistry of Phobos even without actually touching the surface.

These technologies are based on lasers. In conventional labs it is done in special vacuum tanks. A modest laser evaporates a tiny piece of material from the surface of a sample under study. Then the stream of evaporated material is analyzed by an instrument called a mass spectrometer. Could we do the same thing in an open space vacuum between Phobos and the spacecraft? How big should the laser be in order to evaporate a tiny piece of material from the surface of Phobos? Clearly we were unable to launch an SDI-class laser, even if we had it. But fortunately, the final assembly was millions of times below the limits of the ABM treaty.

We were excited at the idea of using this technology, and later on we added another option that would complement the laser. A similar effect could be obtained with an ion beam from a rather modest piece of equipment on board: the so-called ion gun. The ions fired from such a gun in turn had to sputter a small part of the surface material, which would then be subjected to subsequent chemical analysis.

The actual instrumentation for these exotic ideas was eventually built by a large international consortium led by my institute. I myself coauthored a couple of patents on the inventions involved in the experiment. The Czechs contributed an important subsystem, the laser range tracker. There were also parts designed by the Bulgarians, the Finns, and the Germans. Within our country several institutions contributed to the hardware. It was probably one of the most sophisticated experiments I have ever been involved with.

We at the institute were so busy with such instruments that we did not pay any attention to the political maneuvers around the project. On the insistence of the Ministry of General Machine Building, the organizational format of the mission was changed. The scientific team lost its role of supervisor of the whole project, a role it had successfully played on *VEGA*. Apparently our industry contractors were irritated not only by the intervention of scientists in technical and engineering issues, but also with the credit we had received on our leadership role in the *VEGA* mission. Now they wanted to become the chief beneficiaries of recognition—and all that goes with it—if Phobos were to succeed.

Two identical spacecraft carrying our priceless scientific payload were finally launched from Baikonur Cosmodrome in July 1988. By the end of summer we already had a lot of scientific data from instruments assigned to study particles and radiation in interplanetary space. Everything on board the twin spacecraft seemed normal.

Then suddenly, disaster struck out of the blue. On September 2, I got a telephone call from Kremnev, who was serving, on behalf of Glavcosmos, as technical director of the project. "I think we have lost *Phobos I*," he told me grimly.

Apparently a few days earlier, the controllers had sent a very long message to the spacecraft. This consisted of a list of different commands and instructions that the spacecraft had to implement during the brief interruption in communication with Earth—at a time when it had to be controlled by its own on-board computer. One of the specific lines of this software message contained a terrible error. One number—the last digit—was simply omitted. Due to a most unfortunate coincidence, it was the equivalent to a command to literally "commit suicide"—to close the thrusters of the orientation system. The spacecraft, in deep space, was eventually doomed to lose its orientation like a dazed acrobat.

The controllers had made a grave error. There had been no cross-checks. Nobody bothered to analyze the accuracy of the message, which routinely should be done on a special computer simulating the on-board system. On top of everything, the spacecraft itself had to be redundant. Its own electronic brain, the on-board computer, should have immediately rejected such self-destructive instructions.

Since 1973, when we unsuccessfully tried to send a group of spacecraft to Mars, there had been no single failure of this caliber in our deep space program. Everything we had done, sending the space cavalcades to Venus and at last to the comet, created a dangerous kind of comfort—very similar to what probably happened to the controllers of the Chernobyl reactor. Our success had become self-deceptive. The Phobos project consumed 200 million rubles, but fortunately we had the second spacecraft that provided redundancy for the whole mission. Hence, the highest priority was to keep *Phobos II* alive.

The error had emanated from one of IKI's contractors in the military industry. And on discovering what had happened, the bosses in the Ministry of General Machine Building wanted to punish the guilty parties immediately. The controllers of the software team from these industrial contractors were to become the scapegoats—precisely like the aftermath of Chernobyl. Big Oleg called for an urgent meeting in the Central

Committee. Even we scientists, who suffered most from the loss of the spacecraft, were not spared by the infuriated boss.

In a state of utmost irritation, Big Oleg said, "You in the academy are probably happy now that the aerospace industry has made such an error."

No one, of course, was happy. It was most painful to sit through the final assessment of the situation at the meeting of the ministry's board in October 1988. There I urged the ministry officials not to fire the software experts who had made the error, at least not during the life of the project. Exhausting all of the arguments of substance—don't switch horses in the middle of the stream, and so on—I figured out that even the red tape bureaucrats and big bosses would understand a joke better: "As Lavrenti Beria used to say in similar situations, 'Lets make them work; we can shoot them later.'"

I continued: "The controllers made a terrible error, but now we have to save *Phobos II*. The situation is like what happens in a war: they should consider themselves sent to the 'punishment battalion.'"

Now *this* was logic the ministry could understand.

As we all came to painfully recognize, even a layperson's analysis of the *Phobos I* situation could see how easy it would have been to avoid this human error. The system controlling spacecraft operations needed only a few simple precautions against any command like "Please commit suicide."

As a result of the error, the altitude control system was switched off. The spacecraft was left alone for a couple of days and gradually started to lose orientation. It behaved like a free body under the influence of different external forces, all of which were extremely weak. The most important force influencing the spacecraft—the pressure of solar light—acted on the panels as if they were sails. The design was such that within a few days this solar pressure overturned the spacecraft. In the final configuration, the solar panels lost the sun's illumination and the spacecraft was left short of electric power.

If Gorbachev had wanted to use the space program as a role model, the loss of *Phobos I* was a serious blow. However, I hoped that at least direct cheating and corruption were excluded from what we were doing. But neither I nor Gorbachev knew the real extent of what Big Oleg was willing to do.

The next important landmark, to put *Phobos II* in orbit around Mars, was the well-defined moment when the on-board computer issued the command to switch on the engine for the spacecraft to maneuver. I was waiting impatiently for the result of what was normally a rather routine operation. Kremnev's phone message at the end of January 1989 confirmed "we are in orbit [around Mars]." But the tone of his voice was far from victorious: "Roald, we have a real problem with the computer. I am afraid it hasn't enough resources to make its encounter with Phobos."

Apparently out of three identical processors thought to provide enhanced redundancy, one had already died and the second was indicating occasional malfunctioning. What was most frightening was that no commands could be issued if only one processor were to survive. The system's logic was based on a "vote" by each processor, with the majority opinion adopted as the final decision. The remaining sole processor, even if it were in perfect condition, could do nothing. It would be unable to overwhelm the "no" vote of two "dead souls." Unfortunately, this logic could not be changed. The only thing we could do was to pray.

Phobos II was approaching its target. Though we knew perfectly well that the first spacecraft had been lost not simply because of human negligence, we looked optimistically to the prospects for *Phobos II.* All precautions were made within the limits of what we could do on Earth, separated by tens of millions of miles from the spacecraft.

We tried to be extremely careful with all the operations. In February several maneuvers controlled from Earth placed the spacecraft in orbit roughly coinciding with the orbit of Phobos. The mission was close to its culmination, with the date for the final encounter with Phobos chosen for April 7.

During the last few days of March, everyone was busy checking the instruments on board from the data we were receiving. We were anxious to see if the instruments were ready for the final encounter. In these routine communications with the spacecraft, we checked its readiness and got excited. The laser system was in perfect condition.

One of my colleagues, Konstantin Gringaus, during this same period came up with a draft of an article. The space program, according to him, was in great danger. The general public believed the money for expensive space exploration was being unjustly wasted. During the electoral campaign to the Congress of People's Deputies, many candidates, including Yeltsin, openly argued in favor of cutting the space program severely to bring relief to the economy. At the same time, the space establishment resisted any kind of perestroika in its approach to its projects. Gringaus's article questioned how to bring these two issues together and at the same time defend the space program against adversaries who used political rhetoric without substance.

In the article, Gringaus's real criticism was directed toward the space barons. Articles of this type had not been written up to this point in the era of glasnost. That is why it took a lot of effort to publish such a critical article in the "big press." The space industry had a very strong hand in newspaper circles. It controlled many journalists; the restive ones would simply be denied visits to the launch sites. Fortunately, the science editor of *Pravda* took Gringaus's article very seriously. This type of scenario was all quite familiar to him. He had encountered problems of this type only recently when he was sent to Chernobyl as a correspondent. Though there were no direct analogies between the two stories, some parallels were obvious. We did not have to wait long for a reaction from the space establishment after the article was published.

On the very next day Galeev, my successor, went to the control center for one of the last communications with *Phobos II* before the planned encounter. That very evening the spacecraft had to make a new series of images of the Martian moon Phobos at rather close range. To make such images of Phobos, it had to turn toward the target. During such maneuvers, with instruments looking in the direction of Phobos, communications with Earth were interrupted. Each time the program of imaging and measurements was completed, the spacecraft computer had to restore its original orientation and consequently its communications with Earth.

While waiting for the recovery of the signal from the spacecraft, Galeev was surrounded by angry directors from Glavcosmos, attacking him for the article in *Pravda*. They were bewildered by Gringaus's statements, comparing the mismanagement of *Phobos I* with the mismanagement of the *Admiral Nakhimov*. But even more, they were angered by the criticism of the Soviet shuttle *Buran* project. Gringaus had said he couldn't understand why now, when the country needed resources and money, we should copy the American shuttle.

Galeev was under very strong pressure. There were even threats: "If you guys behave this way, we will simply stop providing you with spacecraft altogether."

Galeev responded with firmness: he said he agreed with the substance of Gringaus's article, but that the form and layout of the material were the responsibility of the author. The "execution" of Galeev—the new director of IKI—had to be postponed when the warning of "one-minute readiness" for the reopening of communications with the spacecraft came.

Everyone sat in quiet anticipation. However, at the predicted time for communication, there was silence; no signal came from deep space. It was late in the evening on March 27, only ten days before the scheduled encounter with Phobos. It was late, but no one left the control room. Everyone stayed overnight trying desperately to restore communications with the spacecraft. But there was only one outburst of activity detected from the spacecraft. And then it was gone: exactly four hours later, there was a very weak, unintelligible signal sent by the spacecraft, indicating that the spacecraft was in a state of uncontrolled precession. It was the last message from the dying *Phobos II*.

No one can describe the state of shock in the Space Research Institute—especially among the young men whose only activity throughout their short careers had been to prepare the experimental devices flown on the Phobos mission. They had been extremely proud to be part of one of the most ambitious and unparalleled scientific projects of the space age. But now with the news of the loss, many of them told me that they felt a tremendous sense of emptiness. The older generation of space scientists and engineers know better the risks of their profession. They also know, psychologically, how to survive such failures.

On the official side, representatives from the space industry, Glavcosmos, tried to dampen discussion of the failure by talking about how much scientific knowledge had been gained by the Phobos mission. Out of three goals identified by the mission, two were fulfilled. One was en route science: the study of the sun, solar wind, and the interplanetary medium. Indeed, those studies were rather successful. One particularly important experiment, conducted for the first time in space, charted solar oscillations.[1]

The second goal, the orbital study of Mars, also produced a reasonably good remote scientific study of the Martian atmosphere and surface.

What remained unfulfilled was the work scheduled to commence with the encounter.

On top of dealing with difficult interpretations regarding the loss of spacecraft, there remained one very important issue. The project had

1. This young branch of solar astrophysics, which studies the tremors given off by the sun, is called helioseismology.

involved scientists and engineers from thirteen nations. At least some of them, whose experiments did not get a chance to operate before the mission's sudden termination, were justifiably upset. Their space organizations had expended a lot of effort to get support from their respective national taxpayers. The important issue of the credibility of our space technology would be at stake if our technology were again to be included as part of a platform for a joint international mission.

The Western press definitely gave the opponents of cooperation a chance to raise their voices of doubt about the future of joint missions. I made the suggestion that an investigation of what happened, including every detail, should be conducted jointly with all foreign participants on the project. Only the truth, explored with openness and honesty, could restore the credibility of our program. Without this examination there would be no more opportunities to enjoy broad international cooperation—at least for the foreseeable future.

The aerospace industry, closed to such truth-seeking since its inception, had a well-developed psychological resistance to scrutiny. Its bosses had not only become accustomed to unnecessary secrecy in which they tried to hide actual happenings, they had also learned how to use tricks to avoid responsibility and punishment in the face of failure. The loss of *Phobos II* proved to be no exception.

At IKI we underwent a very painful self-castigating analysis. No one can simply remove a part of the guilt from himself. It is true that we were the customers; nevertheless, we should have been much more insistent when dealing with the aerospace industry. What happened was obvious contributory negligence from our side. In 1986 and 1987 during the most critical period of preparation, I was more and more absorbed by political activities, including arms control assessments and the summits. Somehow I had underestimated the importance of fighting internal "wars" instead of external ones. Too much energy had gone toward arms control and to the fight against forces that wanted to establish a Soviet SDI system. Not only I, but the institute as a whole, had been filled with euphoria over the success of the *VEGA* mission as well as the flight of high-energy scientific telescopes on board the station *Mir*—the first to detect hard radiation from the famous supernova in 1987. We had lost some of our vigilance.

The difference between the space science community and the space industry community rests on the fact that while industry instinctively prefers contracts that repeat projects and models already in existence, scientists need novelty. Old mundane results have no real scientific value. Our profession by definition requires us to move on to new designs. The difference between space science and technology is, essentially, a philosophical conflict between two ways of life—especially in our country, where closeness stimulated a very cautious approach from the industry.

On the eve of the open review process, our industrial contractors tried to neutralize us: "We should first think about the future of the space program. And only then about what portion of the truth should be disseminated to the public. If we decide that the well-being of the space program is more important, than we can try to find a flexible approach: 'Look, a lot of scientific data was already received. It's an outstanding project; after all, only a tiny piece of the last ten days was not fulfilled. But it happens.' Moreover, we could hint at the environmental hostility around Phobos."

The mentality of the closed system justified small lies for a noble cause, in this case the well-being of the space program. From the same enterprises that were taking part in the Phobos project, the younger generation of space engineers were our allies. Afraid that the space elite would try to hide the actual happenings, they would say, "Look, we can show you more than half a dozen very evident technical design errors in the control system." They were ready to submit an independent analysis that they had made to the authorities. And they wanted very much for IKI to support them in the process of such analyses. It was an interplay of how glasnost and democracy confronted the old approach.

The establishment in the space industry knew well that even if a single case were to be opened for general discussion, it would immediately trigger a chain reaction, a reconsideration of everything they controlled. They were also afraid that it would inspire questions such as, Why are such terribly expensive projects as the Soviet shuttle, orbital stations, and superlaunchers still going on?

To silence us in the forthcoming evaluations, industry promised: "Okay, if you help us find the best presentation, the best cover for the whole story, we will do anything for you. For example, we are ready to consider another mission to Phobos as soon as possible—if 1990 is too soon, let it be 1992."

Quite understandably a few people, especially those whose efforts in the last few years had gone awry, wanted such a repetition. But at the same time a sober voice reminded me that we should never make a decision without going all the way to the end in disclosing what had happened. It was a crucial test for me and the institute. We had passed some of our earlier tests with what I thought had been some maturity, and we had had several good projects. Then we made a very important step forward. We extended hospitality to the foreign scientific community and proved that such cooperation was not a Tower of Babel, but that it works. Now we had to pass one more test, the most serious exam. We had to prove that we were decent people who could face failure with a readiness to confess it and learn from it. We had to prove that we were ready for real perestroika. We also had to prove that we were honest with our taxpayers.

The next and even more difficult step was to invite our partners as legal participants in the discussion. The whole multinational team of Phobos was already in Moscow in mid-May 1989 when officials from the government and space industry, including the minister of the Ministry of General Machine Building, came to the presidium of the academy to decide how much should be told at the international gathering.

I seized the chance to make a statement: "This is the most important challenge now. If we are unable to meet it, we will jeopardize our international cooperation. Before we would fly any international project in the future, we have to answer the question, What happened to *Phobos*? If not, we would be unable to persuade our partners that we are going to improve our performance."

As the final result of this meeting, the decision was made to send all the responsible people—the leaders of the space industry involved in the design and integration of the *Phobos* spacecraft—to meet our foreign partners.[2]

On the next day, the biggest room of IKI was filled with 200 scientists—members of the Phobos team, including the international group. The delegates from industry surrounded me at the podium. Two of them had the highest titles of chief designer and general director—the equivalent of being chairman, president, and CEO at the same time: V. Kovtunenko, our principal contractor, and Vladimir Lapygin, czar of the huge institution of automation, the biggest military-industrial enterprise

2. I felt that the minister, Vitaly Doguzhiev, was quite reasonable. He was a practitioner who had had the ministerial chair for only about a year—since the promotion of "Big Oleg" Baklanov. Unfortunately, his responsibility for the space industry soon was over. He was succeeded by "Little Oleg" Shishkin.

in our Moscow district. It was he who was responsible for the control system of the *Phobos* spacecraft.

The first person to speak was my colleague namesake, Roald Kremnev. His boss, Kovtunenko, preferred to hide; Kremnev was considered much more articulate in doing such tricky jobs. Instead of going to the business directly, Kremnev launched into a long speech, speculating on what could have happened. Spacecraft could encounter vicious dust particles, or even meteorites capable of destroying the spacecraft. He hinted that the scientific team had not stressed enough the potential "environmental risks" associated with flying to Mars. Consequently, the spacecraft designers had not taken these risks into account in the design. Second, he said, the spacecraft could be damaged by solar flares, the big outbursts of solar activity.

I looked around the room to the other scientists, and I could see that they were on the verge of laughing at all these explanations. The manipulative quality of Kremnev's presentation was utterly transparent to everyone. But we knew he was "on duty," on behalf of the space industry, trying to cast a fog over the issue and avoiding any hint of computer failure.

Only one excuse was indeed legitimate. The final approval of the budget for the Phobos project had come too late, only in early 1985, which left too tight a schedule for a sophisticated space project. Even with money at hand, without formal approval from the highest governmental level of the Comission on Military-Industrial Issues (VPK), it was very difficult to impose a discipline among contractors. In an economy based on the shortage of everything, every contractor was *over*-contracted. Among different contracts, first priority would be given to those confirmed by the highest existing authority: the Central Committee, the Council of Ministers, the VPK.

Despite all of my irritation and frustration, at the end of the discussion I could not resist a bit of ridiculing. I told the gathering that I was ready to accept an interpretation involving a collision with an unidentified extraterrestrial body, if only Kremnev would agree that the body was the Loch Ness monster.

The actual danger from dust particles can be precalculated. We had done such an analysis in the institute and found that the probability of encounter was extremely low. Besides, we had asked the Jet Propulsion Laboratory (JPL) in Pasadena to do an independent assessment based on the long exposure of the *Viking* spacecraft, which had been in essentially the same area. The assessment coincided with ours.

Under cross-interrogation by the scientists, the industrialists confessed that after they recovered communications with the *Phobos II* spacecraft for a brief moment, they discovered that the spacecraft was spinning around the wrong axis. In such a configuration the influx of solar energy was too low to power the craft's electric battery, and so the mission was already doomed. It was clear for everyone that independent of what cause had led to the sudden loss of orientation, the spacecraft did not have enough resources of "survivability."

I was waiting for contributions from our foreign colleagues. My old comrade-in-arms in planetary exploration, Bruce Murray, former director of JPL, said that it was shameful how the disaster happened. There was an obvious lack of a systematic approach, bringing together science and industry. In all future projects, the science team including its international component would have to be a part of systems design and planning.

Murray's eloquent speech, supported by Jacques Blamont on behalf of the French Space Agency, and by a few others, provided an impressive show of international solidarity of the "proletarians of intellectual labor" against the corrupt barons of the Soviet space mafia.

Finally, I suggested that there was only one way to restore the credibility of Soviet space technology—to set up an international commission and to give an opportunity to all the participants, including the international team, for access to all the details of the Phobos project failure. Even Aleksandr Dunaev, the head of Glavcosmos and a genuine personification of an apparatchik, had to accept that proposal.

Everyone thought the meeting was a success; we had recovered the format and structure of the work used for the *VEGA* project, and maybe in the future it could even improve. Altogether the direct encounter with "these types" was, I believe, an eye-opener for my foreign friends, who confessed to being utterly astonished at seeing such men from the stone age. "Now we understand why you need perestroika," they said. The space industry had developed a special capability to survive in such an environment, when everything was kept under secrecy. Now they were afraid to start a new life with glasnost. However, the promise of that international gathering on Phobos was never fulfilled. The officials didn't even try to continue the investigation into the causes of the failure.

In Place of an Epilogue

➢

The sad story of the *Phobos* project revisited me in the fall of 1993, when the news came of the loss of the Mars *Observer* probe. I could understand the pain and disappointment of my colleagues from NASA. Alas, no space mission is insured against the risk of technical mishap. The strength of a space community and a nation is tested by the way such calamities are handled. The Soviets tried hard to hide the facts related to failures. But a system unable to learn from its own mistakes must be doomed.

With the military-industrial complex under the growing pressure created by perestroika, it was perhaps a mistake that I did not apply enough weight to ensure the technical investigation of the *Phobos* loss. But I was no longer the director of IKI and, more importantly at that time, I was consumed by my political election to the then-new Soviet parliament, the Congress of People's Deputies. Against considerable resistance from my dedicated enemies in the Academy of Sciences, I was elected, together with Andrei Sakharov.

On May 25, 1989, the Congress of People's Deputies came into session. It was comprised of a strange mix of over 2,000 people. The majority did not differ from the old-style apparatchiks, such as my old nemesis, "Big Oleg" Baklanov who was sent to the congress as a representative of the Communist party.

Among other things, the congress had to elect the first public body to oversee the "complex." To my dismay, instead of going to one of the liberal reformers, the post of chairman went to none other than Vladimir Lapygin, whose mailbox had been responsible for the failure of the onboard control unit of the *Phobos* spacecrafts. The maneuver was quite obvious to me. Big Oleg, the real power behind the scene, desperately needed to keep this post under his control. Who would be a better choice than Lapygin, the old subordinate of Big Oleg, and himself one of the biggest contractors for the military? This is how Lapygin escaped the scrutiny of international cross-examination of the *Phobos* project.

But it was only the last brief triumph of an obsolete system. History had already started its countdown of the last days of the Bolsheviks' "grand historic experiment," which collapsed with the failed putsch of August 1991.

Baklanov was one of the three key government conspirators in that putsch. Gorbachev, impressed by both his entrepreneurial talent and his ideological purity, had given him the job of running the space industry. Perhaps it was precisely this last quality that provoked Baklanov to undertake an act of desperation. As a result, Big Oleg went to prison and Gorbachev's authority was also seriously eroded.

The departure of Gorbachev from the political scene left Yeltsin face to face with this huge military-industrial apparatus. Would "Yeltsin's revolution" overcome its iron embrace?

While Yeltsin was winning the nation and the world's sympathy, he used attractive populist slogans. With a pacifist's simplicity he called an immediate halt to a wide array of military programs, as if he wanted to become a greater champion of disarmament than Gorbachev.

Meanwhile, the complex realigned after the failure of the coup. Warning of the consequences of quick reforms and the danger of social unrest that would be provoked by massive layoffs from military enterprises, they put Yeltsin under relentless pressure. The new Russian president compromised. The green light was given for the complex to again actively pursue sales of arms in the international marketplace. Leading mailboxes were given extended special subsidies to prevent bankruptcy. There was and has been an active ongoing campaign for the revival of Russia as a military superpower.

While the military-industrial machine has already demonstrated that it will outlive the system that created it, Soviet science has not been so lucky. The fabric of intellectual life in the former Soviet Union has in large part been destroyed under the collapsing ruins of the old regime.

My mentors, the generation of Kapitsa and Leontovich, were capable of enduring the long dark period of totalitarian indoctrination. They died before the end of Soviet communism so, in that sense, they did not live to enjoy the dividends of freedom.

Now it is for the generation of my contemporaries and their pupils to get through the economic cataclysms of the postcommunist transition.

The "unmaking of the Soviet scientist" as a destructive process has already taken place. A part of the former intellectual elite—the ones who were recognized in the international scientific community—left, constituting a regular brain drain. Many from the younger generations were removed from the research institutes by a wave of "internal brain drain." They are now struggling to survive outside the realm of science—the world that was an escape and asylum for the Soviet scientists of my time.

But many, despite the pressures of mundane life, stay firm in their selfless service to science. God help them to do so with the same grace, tenacity, and integrity that distinguished that special breed of scientists, "the keepers of the flame," that were Kapitsa and Landau, Leontovich and Sakharov.

Glossaries

➢

Glossary of Russian and Soviet Names

(Editor's Note: The intent of this glossary is not to include all the individuals in the book, but to give the reader a quick reference to those that appear in more than one instance.)

Sergei Afanasiev (aka "the Big Hammer") Minister of General Machine Building.

Marshal Segei Akhromeyev Chief of general staff.

Anatoly Alexandrov Keldysh's successor as president of the Soviet Academy of Sciences.

Georgy Arbatov Political scientist and advisor to Gorbachev.

Lev Artsimovich Soviet scientist and physicist who led the fusion program.

Georgy Babakin Soviet designer of Lunnik and Lunokhod spacecrafts.

Oleg Baklanov (aka "Big Oleg") Leader of the military-industrial complex under Gorbachev.

Andrei Budker Soviet physicist and the founder of the Institute of Nuclear Physics.

Konstatin Bushuyev Soviet director of the *Apollo-Soyuz* project.

Vladimir Chelomey Soviet rocket designer.

Colonel Georgy Chernyshov Security commissar at IKI.

Yuri Gagarin Soviet cosmonaut who was the first man in space.

Alec Galeev Successor to the author at IKI.

Valentin Glushko Soviet space luminary.

Georgy Grechko Soviet cosmonaut.

Konstantin Gringaus Soviet space scientist at IKI.

Peter Kapitsa Soviet physicist who established the Institute of Physical Problems in the mid-1930s; awarded the Nobel Prize in Physics in 1978.

Mstislav Keldysh "Chief theorist of cosmonautics" and the president of the Soviet Academy of Sciences.

Yuli Khariton Soviet nuclear weapons designer.

Vladimir Kirillin Member of the Soviet Academy of Sciences and later deputy prime minister to Alexei Kosygin.

Sergei Korolev Soviet space luminary and the founder of the Energiya Design Bureau; considered to be "father" of Sputnik.

Igor Kurchatov Soviet physicist who was the leader of the Soviet nuclear program in the 1940s and 1950s.

Lev Landau Pupil of Niels Bohr and a star of Soviet theoretical physics; awarded the Nobel Prize in Physics in 1962.

Vladimir Lapygin Soviet spacecraft designer.

Mikhail Lavrentiev Soviet hydrodynamicist and the founder of Akademgorodok.

Oleg Lavrentiev Soviet soldier who invented the electric confinement of hot plasma.

Mikhail Leontovich Soviet theoretical physicist who was the head of Kurchatov Plasma Theory.

Yevgeny Lifshitz Soviet theoretical physicist.

Mikhail Lomonosov Great Russian scientist of the 18th century.

Trofym Lysenko Soviet biologist who was the chief modern proponent of the idea of the inheritance of acquired characteristics, in opposition to mainstream biology and genetics.

Guri Marchuk Successor to Alexandrov as president of the Soviet Academy of Sciences.

Vasily Mishin Soviet spacecraft designer and the successor to Korolev.

Major General Georgy Narimanov Soviet expert in fluid dynamics of rocket fuel tanks and the deputy director of IKI.

Georgy Petrov Soviet academician who preceded the author as director of IKI.

Bruno Pontecorvo Italian nuclear physicist who emigrated to England and then defected to the Soviet Union.

Aleksandr Prokhorov Soviet scientist who was awarded the Nobel Prize in Physics in 1964.

Yuri Rumer Soviet theoretical physicist, arrested during the Great Purges.

Andrei Sakharov Soviet physicist who designed the H-bomb and later became an opponent of the Communist regime.

Ivan Serbin Head of the Defense Department of the Central Committee of the CPSU.

Nikolai Semenov Soviet chemist who was awarded the Nobel Prize in Chemistry in 1956.

Yuri Semenov Soviet spacecraft designer.

Vitaly Shafranov Soviet plasma theorist and the author's colleague at the Kurchatov Institute.

Aleksandr Sheindlin Soviet scientist who founded the Institute of High Temperatures.

Igor Tamm Soviet theoretical physicist who was awarded the Nobel Prize in Physics in 1958.

Gherman Titov Soviet cosmonaut.

Konstantin Tsiolkovsky Russian inventor of early rockets; considered to be "father" of Soviet "cosmonautics."

Marshal Dmitri Ustinov (aka "Uncle Mitya") Secretary of the Central Committee of the CPSU, and later, minister of defense.

Glossary of Key Selected Agencies and Institutions

AgroProm Acronym of the agricultural-industrial complex.

Akademgorodok ("Academic City") New center for science in Novosibirsk.

All-Union Research Institute of Experimental Physics (aka Moscow Center 300, Near Volga Office, and Aramaz-16) Official name of Sakharov's installation.

Baikonur Cosmodrome Soviet space launch site; now on the territory of Kazakhstan.

Bureau of Electronic Equipment Area of the Kurchatov Institute where author was assigned—renamed the Division of Plasma Research.

Commission on Military-Industrial Issues (VPK) Part of the Council of Ministers.

Division of Plasma Research New name given to the Bureau of Electronic Equipment.

Glavcosmos Agency Instrument created to promote commercial services of the Soviet space program.

Gosplan State Planning Committee.

IKI Acronym of the Space Research Institute.

Institute of Applied Mathematics Establishment directed by Keldysh where computational techniques toward strategic applications were developed.

Institute of High Temperature Establishment where author headed the plasma theory lab.

Institute of Nuclear Physics Budker's institute in Akademgorodok.

Institute of Physical Problems Kapitsa's institute in Moscow.

Intercosmos Agency created to promote international cooperation in space.

Kurchatov Institute of Atomic Energy (aka Laboratory of Measuring Instruments) Official name of Kurchatov's installation.

Lavochkin Design Bureau Aerospace company that contracted for deep space probes.

Ministry of General Machine Building Overseer of the space industry.

Ministry of Medium Machine Building Overseer of nuclear programs.

Moscow Energy Institute Institution where author began teaching in 1959.

Physico-Technical Institute The chief cradle of Soviet fundamental and applied physics located in Leningrad.

Red Star Company where space nuclear reactors were built.

Space Research Institute (IKI) Representative of the Soviet Academy of Sciences in the space program.

State Committee on Science and Technology Headquarters of the science and engineering establishment.

VPK Acronym of the Commission on Military-Industrial Issues.

Index

➢